Ceramics for Environmental and Energy Applications

Ceramics for Environmental and Energy Applications

Ceramic Transactions, Volume 217

A Collection of Papers Presented at the 8th Pacific Rim Conference on Ceramic and Glass Technology
May 31–June 5, 2009
Vancouver, British Columbia

Edited by
Aldo Boccaccini
James Marra
Fatih Dogan
Hua-Tay Lin
Toshiya Watanabe

Volume Editor
Mrityunjay Singh

The American Ceramic Society

WILEY

A John Wiley & Sons, Inc., Publication

Published by John Wiley & Sons, Inc., Hoboken, New Jersey.
Published simultaneously in Canada.

Library of Congress Cataloging-in-Publication Data is available.

ISBN 978-0-470-90547-0

Printed in the United States of America.

10 9 8 7 6 5 4 3 2 1

Contents

SOLID OXIDE FUEL CELLS AND HYDROGEN TECHNOLOGY

CERAMICS FOR ELECTRIC ENERGY GENERATION, STORAGE AND DISTRIBUTION

PHOTOCATALYTIC MATERIALS

DIRECT THERMAL TO ELECTRICAL ENERGY CONVERSION MATERIALS

Preface

This volume is a collection of 17 papers from five symposia held during the 8th Pacific Rim Conference on Ceramic and Glass Technology (PACRIM-8) in Vancouver, British Columbia, Canada, May 31–June 5, 2009. These symposia include:

- Glasses and Ceramics for Nuclear and Hazardous Waste Treatment
- Solid Oxide Fuel Cells and Hydrogen Technology
- Ceramics for Electric Energy Generation, Storage and Distribution
- Photocatalytic Materials: Reaction, Processing, and Applications
- Direct Thermal to Electrical Energy Conversion Materials and Applications

PACRIM-8 is the eighth in a series of international conferences on emerging ceramic technologies that began in 1993. PACRIM-8 was organized and sponsored by The American Ceramic Society and was endorsed by The Australian Ceramic Society, The Ceramic Society of Japan, The Chinese Ceramic Society, and The Korean Ceramic Society. A total of 862 experts, practitioners, and end users from forty-one countries attended PACRIM-8, making it one of the most successful ceramic science and engineering events in recent years.

We would like to thank The American Ceramic Society for organizing and sponsoring the meeting and for making this proceedings volume possible. We also thank the authors, participants, and reviewers of this PACRIM proceedings issue. We hope that this issue becomes a useful resource that contributes to the overall advancement of materials in environmental and energy applications.

ALDO BOCCACCINI
JAMES MARRA
FATIH DOGAN
HUA-TAY LIN
TOSHIYA WATANABE

Introduction

The 8th Pacific Rim Conference on Ceramic and Glass Technology (PACRIM 8), was the eighth in a series of international conferences that provided a forum for presentations and information exchange on the latest emerging ceramic and glass technologies. The conference series began in 1993 and has been organized in USA, Korea, Japan, China, and Canada. PACRIM 8 was held in Vancouver, British Columbia, Canada, May 31–June 5, 2009 and was organized and sponsored by The American Ceramic Society. Over the years, PACRIM conferences have established a strong reputation for the state-of-the-art presentations and information exchange on the latest emerging ceramic and glass technologies. They have facilitated global dialogue and discussion with leading world experts.

The technical program of PACRIM 8 covered wide ranging topics and identified global challenges and opportunities for various ceramic technologies. The goal of the program was also to generate important discussion on where the particular field is heading on a global scale. It provided a forum for knowledge sharing and to make new contacts with peers from different continents.

The program also consisted of meetings of the International Commission on Glass (ICG), and the Glass and Optical Materials and Basic Science divisions of The American Ceramic Society. In addition, the International Fulrath Symposium on the role of new ceramic technologies for sustainable society was also held. The technical program consisted of more than 900 presentations from 41 different countries. A selected group of peer reviewed papers have been compiled into seven volumes of The American Ceramic Society's Ceramic Transactions series (Volumes 212-218) as outlined below:

- **Innovative Processing and Manufacturing of Advanced Ceramics and Composites, Ceramic Transactions, Vol. 212,** Zuhair Munir, Tatsuki Ohji, and Yuji Hotta, Editors; Mrityunjay Singh, Volume Editor
  Topics in this volume include Synthesis and Processing by the Spark Plasma

Method; Novel, Green, and Strategic Processing; and Advanced Powder Processing

- **Advances in Polymer Derived Ceramics and Composites, Ceramic Transactions, Vol. 213,** Paolo Colombo and Rishi Raj, Editors; Mrityunjay Singh, Volume Editor
  This volume includes papers on polymer derived fibers, composites, functionally graded materials, coatings, nanowires, porous components, membranes, and more.
- **Nanostructured Materials and Systems, Ceramic Transactions, Vol. 214,** Sanjay Mathur and Hao Shen, Editors; Mrityunjay Singh, Volume Editor
  Includes papers on the latest developments related to synthesis, processing and manufacturing technologies of nanoscale materials and systems including one-dimensional nanostructures, nanoparticle-based composites, electrospinning of nanofibers, functional thin films, ceramic membranes, bioactive materials and self-assembled functional nanostructures and nanodevices.
- **Design, Development, and Applications of Engineering Ceramics and Composite Systems, Ceramic Transactions, Vol. 215,** Dileep Singh, Dongming Zhu, and Yanchun Zhou; Mrityunjay Singh, Volume Editor
  Includes papers on design, processing and application of a wide variety of materials ranging from SiC SiAlON, ZrO_2, fiber reinforced composites; thermal/environmental barrier coatings; functionally gradient materials; and geopolymers.
- **Advances in Multifunctional Materials and Systems, Ceramic Transactions, Vol. 216,** Jun Akedo, Hitoshi Ohsato, and Takeshi Shimada, Editors; Mrityunjay Singh, Volume Editor
  Topics dealing with advanced electroceramics including multilayer capacitors; ferroelectric memory devices; ferrite circulators and isolators; varistors; piezoelectrics; and microwave dielectrics are included.
- **Ceramics for Environmental and Energy Systems, Ceramic Transactions, Vol. 217,** Aldo Boccaccini, James Marra, Fatih Dogan, Hua-Tay Lin, and Toshiya Watanabe, Editors; Mrityunjay Singh, Volume Editor
  This volume includes selected papers from four symposia: Glasses and Ceramics for Nuclear and Hazardous Waste Treatment; Solid Oxide Fuel Cells and Hydrogen Technology; Ceramics for Electric Energy Generation, Storage, and Distribution; and Photocatalytic Materials.
- **Advances in Bioceramics and Biotechnologies, Ceramic Transactions, Vol. 218;** Roger Narayan and Joanna McKittrick, Editors; Mrityunjay Singh, Volume Editor
  Includes selected papers from two cutting edge symposia: Nano-Biotechnology and Ceramics in Biomedical Applications and Advances in Biomineralized Ceramics, Bioceramics, and Bioinspiried Designs.

I would like to express my sincere thanks to Greg Geiger, Technical Content Manager of The American Ceramic Society for his hard work and tireless efforts in the publication of this series. I would also like to thank all the contributors, editors, and reviewers for their efforts.

MRITYUNJAY SINGH
Volume Editor and Chairman, PACRIM-8
Ohio Aerospace Institute
Cleveland, OH (USA)

Glasses and Ceramics for Nuclear and Hazardous Waste Treatment

DEVELOPMENT OF GLASS COMPOSITIONS TO IMMOBILIZE ALKALI, ALKALINE EARTH, LANTHANIDE AND TRANSITION METAL FISSION PRODUCTS FROM NUCLEAR FUEL REPROCESSING

J.C. Marra and A.L. Billings
Savannah River National Laboratory, Aiken, SC, USA

J.V. Crum, J.V. Ryan, and J.D. Vienna
Pacific Northwest National Laboratory, Richland, WA, USA

ABSTRACT

The Advanced Fuel Cycle Initiative (AFCI) waste management strategy revolves around specific treatment of individual or groups of separated waste streams. A goal for the separations processes is to efficiently manage the waste to be dispositioned as high level radioactive waste. The Advanced Fuel Cycle Initiative (AFCI) baseline technology for immobilization of the lanthanide (Ln) and transition metal fission product (TM) wastes is vitrification into a borosilicate glass. A current interest is to evaluate the feasibility of vitrifying combined waste streams to most cost effectively immobilize the wastes resulting from aqueous fuel reprocessing.

Studies showed that high waste loadings are achievable for the Ln only (Option 1) stream. Waste loadings in excess of 60 wt % (on a calcined oxide basis) were demonstrated via a lanthanide borosilicate (LaBS) glass. The resulting glasses had excellent relative durability as determined by the Product Consistency Test (PCT). For a combined Ln and TM waste stream glass (Option 2), noble metal solubility was found to limit waste loading. However, the measured PCT normalized elemental releases for this glass were at least an order of magnitude below that of Environmental Assessment (EA) glass. Current efforts to evaluate the feasibility of vitrifying combined Ln, TM, alkali (Cs is the primary radionuclide of concern) and alkaline earth (Sr is the primary radionuclide of concern) wastes (Option 3) have shown that these approaches are feasible. However, waste loading limitations with respect to heat load (Cs/Sr loading), molybdenum solubility and/or noble metal solubility will likely be realized and must be considered in determining the cost effectiveness of these approaches.

INTRODUCTION

The United States Department of Energy (DOE) Advanced Fuel Cycle Initiative (AFCI) is designed to demonstrate a proliferation-resistant integrated nuclear fuel cycle. This fuel cycle consumes transuranic (TRU) elements and supports growth of carbon-free international nuclear energy markets. Building on the knowledge gained over the last 60 years of nuclear science and engineering, the proposed recycling system is not only more sustainable than prior concepts, it will generate less waste and reduce thermal and long-lived radiation impacts on a geologic repository.

The AFCI waste management strategy revolves around specific treatment of individual or groups of separated waste streams. It is a goal for the separations processes to minimize the amount and radiotoxicity of waste to dispose. The waste streams will be separated by either of two set of processes; 1) UREX+ and/or 2) electrochemical. The UREX+ set of processes are nitric acid based solvent extraction processes. The electrochemical set of processes are performed in a chloride based molten salt and include dissolution of the less noble metals followed by electro-reduction.

An example UREX+ flowsheet is shown schematically in figure 1. The example UREX+ flowsheet highlights three (of many) waste streams that require immobilization:

- Separated cesium and strontium waste (Cs/Sr) in the CCD-PEG (or FPEX) raffinate, also contains barium and rubidium

- The raffinate from transuranic extraction (TRUEX) referred to as the transition metal fission product (TM) waste stream, and
- Trivalent actinide lanthanide separations by phosphorus-reagent extraction from aqueous complexes (TALSPEAK) product containing lanthanides and yttrium (Ln stream).

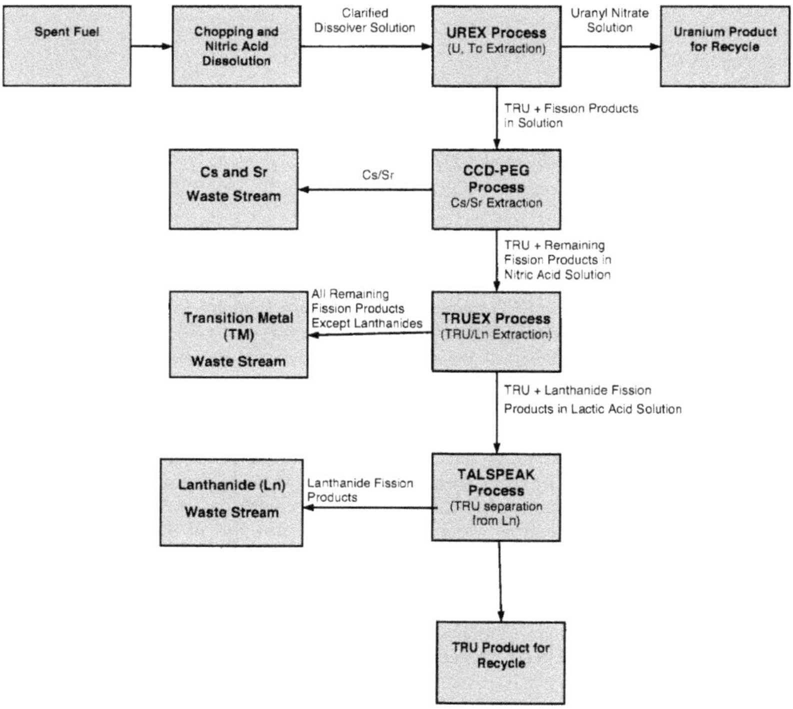

Figure 1. ExampleUREX+ flowheet depicting Cs/Sr, TRUEX and TALSPEAK separations processes and resulting waste streams.

The AFCI baseline technology for immobilization of the Ln/TM wastes is vitrification into a borosilicate glass. Vitrification of the Ln and TM streams have several advantages, including relatively high waste loadings, a proven technology, and similarity in form to waste forms currently accepted for repository disposal. The combination of the separated Cs/Sr wastes and Ln/TM wastes may be a cost effective approach by minimizing the number of waste forms to be fabricated.

EXPERIMENTAL DETAILS

Initial studies concentrated on the development and testing of glasses for the Ln only waste stream (Option 1) and the combined Ln + TM waste stream (Option 2). Recent studies have included the Cs/Sr waste stream to formulate and test a combined Cs/Sr/Ln/TM glass waste form (Option 3).

The testing was conducted in a serial manner where several glass formulations were developed in phases building off of results from the previous phase and incorporation of changes in waste compositions to reflect new knowledge gained from separations flowsheet development and testing. For brevity and illustration purposes, only one or two glass compositions will be discussed for each waste immobilization option.

Waste Compositions

The waste composition of the Ln-only stream is projected to consist of a majority of the lanthanides from the lanthanide series as oxides (Ln_2O_3). The major lanthanide oxides expected to be present are Ce_2O_3, La_2O_3, Nd_2O_3 and Pr_2O_3. Combining the TM stream with the Ln stream adds noble metal fission products (primarily RuO_2, Rh_2O_3 and PdO) to the combined stream. It is expected that a significant amount of ZrO_2 will also be introduced to the combined Ln/TM stream by the introduction of the TM stream.

When the Cs/Sr waste stream is added to Ln/TM combined stream, the main constituents added are Cs_2O, SrO, BaO and MoO_3. The addition of the Cs/Sr stream increases the relative heat load of the waste significantly. Depending on variations in potential flowsheets, the efficiency of the separations processes and the relative partitioning of elements in the waste streams, the combined Cs/Sr/Ln/TM stream was represented by two extremes. One extreme consisted of high MoO_3 content while the other extreme had high noble metal contents. Both MoO_3 and noble metals can have significant impacts on waste loading. It is expected the waste stream compositions will become more definitive as the separations processes mature and are tested. However, at this stage testing extreme levels appeared to be a prudent approach.

Identification of Candidate Glass Forming Systems

Option 1- A lanthanide borosilicate (LaBS) glass was identified as a suitable candidate for immobilization of the lanthanide waste stream, because the waste stream consists mainly of rare earth elements.

Option 2- An alkali borosilicate glass was identified as a suitable candidate for the immobilization of the combined Ln and TM waste stream, because of the complex chemical makeup of the waste.

Option 3 – Alkali borosilicate glass systems were identified as likely candidates for immobilization of the combined Cs/Sr/Ln/TM waste stream, because of the complex chemical makeup of the waste and high variability of MoO_3, noble metals, and ZrO_2.

Glass Fabrication

Each test matrix glass was prepared from the proper proportions of reagent-grade metal oxides, carbonates, H_3BO_3, and salts to produce sufficient glass for the associated testing (typically 200 grams). In general, the raw materials were thoroughly mixed and placed into platinum-alloy crucibles (Pt, Pt/Rh, or Pt/Au). In the case of glasses containing high levels of noble metal oxides (RuO_2, Rh_2O_3, PdO, Ag_2O, etc.), the batch materials were melted in alumina or silica crucibles in order to prevent undesired reactions with the glass and platinum-alloy crucibles during melting. The well mixed batches were placed into electrically heated, high-temperature furnaces at melting temperatures between 1250°C and 1400°C. After an isothermal hold at the targeted melting temperature for nominally 2 hours, the crucibles were removed from the furnaces, and the glasses were poured onto a clean stainless steel plate and allowed to air cool (quenched). In some cases the glass was removed from the furnace and allowed to air cool while remaining in the crucible. The glass pour patties were used as the sampling stock for the various property measurements (e.g. product consistency test [PCT], liquidus temperature [T_L], X-ray diffraction [XRD] and chemical composition).

Durability as Defined by Product Consistency Test (PCT)

The PCT was performed in triplicate on quenched samples of glasses to assess chemical durability using the ASTM C-1285 Method A Procedure [1]. Also included in the PCT testing sets were the Environmental Assessment (EA) glass, the Approved Reference Material (ARM) glass, and blanks. The resulting solutions (leachates) were analyzed via ICP-AES for Si, B, Na, and Li concentrations. Normalized release rates were then calculated based on target and measured compositions using the average of the common logarithms of the leachate concentrations.

Liquidus Temperature (T_L)

The liquidus temperature (T_L) of glasses were determined using an isothermal liquidus determination method [2]. In this method, a glass sample was subjected to a set temperature for nominally 24 hours. The sample was then evaluated using optical microscopy and/or quantitative XRD. For optical microscopy, the sample was thin-sectioned and examined for the presence of crystals. The isothermal heat treatments were continued until the T_L was identified as the temperature between the highest temperature at which a heat-treated sample contained crystals and the lowest temperature without crystals. For XRD, the sample was ground to a fine powder and spiked with known about of CaF_2 (internal standard) to identify crystalline phase type(s) and concentration(s) of the unknown crystalline phase(s) present within the sample. Crystal fraction results were collected for a range of temperatures below T_L. Results are then plotted as mass % crystallinity vs. temperature and a linear fit is used to extrapolate to T_L.

RESULTS AND DISCUSSION

Glass Compositions

The targeted glass compositions that were tested are shown in Table I. Two glasses are discussed for the Ln only waste with the compositions differing significantly in the relative concentrations of Al_2O_3 and B_2O_3. Both compositions resulted in homogeneous glasses with total lanthanide loadings of about 60 wt % (on a calcined oxide basis). Glass Ln-01-2008 was formulated for the Ln only waste stream in the first series of glasses based on LaBS glass data from previous studies [3-5]. Glass Ln-X-SUB was based on a LaBS glass compositions previously developed for actinide disposition [6].

The waste loading for the Ln/TM glass was constrained to result in a total noble metal loading of 2 wt % noble metals in the glass. As noted above, an alkali borosilicate glass was selected for this waste composition.

The Cs/Sr/Ln/TM-1-2.5 glass was formulated with a constraint of a total MoO_3 content of 2.5 wt % in the glass. The Cs/Sr/Ln/TM-2-3.0 glass was limited to 3.0 wt % total noble metal oxide content in the glass. The noble metal content was increased in this composition based on the previously performed Option 2 testing.

Table I. Targeted Glass Compositions for Option 1) Lanthanides, Option 2) Lanthanides + Transition Metal Fission Products, Option 3) Cs + Sr + Lanthanides + Transition Metal Fission Products

	Option 1		Option 2	Option 3	
Component Wt %	**Ln-X-SUB**	**Ln-01-2008**	**Ln+TM-Fe**	**Cs/Sr/Ln/ TM-1-2.5**	**Cs/Sr/Ln/ TM-2-3.0 ***
Ag_2O	-	-	0.03	0.07	0.11
Al_2O_3	8.79	15.75	4.81	5.95	6.53
B_2O_3	11.43	6.50	9.62	5.00	5.16
BaO	-	-	-	1.41	2.20
CaO	-	-	4.81	7.00	5.16
CdO	-	-	0.07	0.07	0.11
Ce_2O_3	13.82	14.20	1.78	1.98	3.09
Cs_2O	-	-	-	1.84	2.87
Eu_2O_3	0.80	0.82	0.10	0.11	0.17
Gd_2O_3	0.94	0.93	0.12	0.10	0.16
La_2O_3	7.05	7.24	0.91	1.01	1.58
Li_2O	-	-	2.41	4.02	3.21
MoO_3	-	-	0.22	2.50	0.78
Na_2O	-	-	8.42	7.00	7.22
Nd_2O_3	23.27	23.91	3.00	3.36	5.22
PdO	-	-	0.80	0.01	0.02
Pr_2O_3	6.43	6.60	0.83	0.93	1.44
Rb_2O	-	-	-	0.27	0.42
Rh_2O_3	-	-	0.19	0.05	0.07
RuO_2	-	-	1.01	0.13	0.18
Sb_2O_3	-	-	0.01	-	-
SeO_2	-	-	0.02	0.05	0.08
SiO_2	17.58	16.10	57.36	53.03	49.94
Sm_2O_3	4.88	5.02	0.63	0.69	1.07
SnO_2	-	-	0.07	0.04	0.07
SrO	2.20	-	-	0.63	0.98
TeO_2	-	-	0.31	0.42	0.65
Y_2O_3	2.81	2.88	0.36	0.40	0.63
ZrO_2	-	-	2.12	1.91	0.87

* Waste loading at 3 wt % noble metal oxide content but noble metal oxides only added at levels where there were expected to be soluble

PCT Results

The PCT results for Option 1 – Ln glass formulations are shown in Table II. Similar to values reported in previous LaBS glass studies, the PCT release values were extremely low (<0.08 g/L) [6-7]. This is compared to the baseline reported value for repository acceptance of EA glass at an average normalized boron release of 16.695 [8]. In almost all cases for Option 1, normalized elemental releases were below the measurement detection limit. PCT results for the Option 2 glass are also reported in Table II. All of the measured PCT normalized elemental releases were also at least an order of magnitude below that of EA, though slightly higher than the Option 1 glasses. PCT results for the Option 3 glasses were comparable and normalized releases were better than an order of magnitude lower than the EA glass for both glasses.

Table II. Measured PCT Response for AFCI Glasses

	Glass ID	Normalized Release (g/L)			
		Li	**B**	**Na**	**Si**
Option 1	LN-X-SUB	N/A	<0.080	N/A	<0.002
	LN-01-2008	N/A	<0.081	N/A	0.003
Option 2	LN+TM-Fe	0.368	0.185	0.089	0.095
Option 3	Cs/Sr/Ln/TM-1-2.5	0.511	0.214	0.420	0.157
	Cs/Sr/Ln/TM-2-3.0	0.486	0.245	0.418	0.149
	EA [11]	9.565	16.695	13.346	3.922

Liquidus Temperature (T_L) Results

The measured T_L and primary crystalline phases of the Option 1 glasses are given in Table III. The optical microscopy method was used for determination of T_L for the option 1 glasses. XRD was used to determine the primary crystalline phases of the Option 1 glasses. Cerianite (LnO_2) structure was identified in LN-01-2008. Lanthanide borosilicate ($Ln_3BSi_2O_{10}$) structure was identified in LN-X-SUB.

The measured T_L and primary crystalline phase for the Option 2 glass are given in Table III. The crystal fraction method (XRD) was used for measurement of T_L for this glass. This switch in method was made because the noble metals are almost entirely undissolved in the glasses, and at the higher concentration of noble metals in this glass made it difficult to determine if other crystalline phases were present near T_L. The undissolved noble metals don't significantly interfere with identification and quantification of other crystalline phases in the crystal fraction method. XRD was used to determine the primary crystalline phase in the glass. The LN+TM-Fe (Fe_2O_3 removed), had a primary crystalline phase of $Ca_{2.2}Nd_{7.8}(SiO_4)_6O_{1.9}$.

The measured T_L and primary crystalline phase for the Option 3 glasses are shown in Table III. These liquidus temperatures would support glass processing at a nominal 1250° C processing temperature. The primary crystalline phase, $Ca_2Nd_8Si_6O_{26}$, was the same for both the MoO_3 and noble metal limited glasses.

Table III. Measured T_L and Primary Crystalline Phases of Option 1 Glasses

	Glass ID	Measured T_L, °C	Primary Crystalline Phase(XRD)
Option 1	LN-X-SUB	1279	$Ln_3BSi_2O_{10}$
	LN-01-2008	1315	Cerianite (LnO_2)
Option 2	LN+TM-Fe	1036 (XRD)	$Ca_{2.2}Nd_{7.8}(SiO_4)_6O_{1.9}$
Option 3	Cs/Sr/Ln/TM-1-2.5	1017	$Ca_2Nd_8Si_6O_{26}$
	Cs/Sr/Ln/TM-2-3.0	1128	$Ca_2Nd_8Si_6O_{26}$

CONCLUSIONS

Glass compositions have been developed for the Ln waste steam alone (Option 1), the combined Ln and TM waste streams (Option 2) and the combined Cs, Sr, Ln and TM (Option 3) waste streams generated by the AFCI UREX+ family of processes.

For the Ln only (Option 1) stream, lanthanide loadings in excess of 60% (on a calcined oxide basis) were demonstrated using a lanthanide borosilicate (LaBS) glass system.

- The glasses had excellent relative durability as determined by the PCT.
- The measured T_L ranged from 1315°C to 1279°C for the Option 1 glasses. These values are well below the nominal 1400°C processing temperature for these glass compositions.

For the Ln + TM glasses (Option 2), noble metal solubility will likely limit waste loading. A glass composition was developed and tested that contained 2 wt % noble metals.

- The measured PCT normalized elemental releases for this glass were at least an order of magnitude below that of EA glass.
- The measured T_L was 1036°C which is sufficiently below the 1250°C processing temperature for this glass.

For the Cs/Sr/Ln/TM glasses (Option 3), noble metal solubility or Mo solubility are expected to limit waste loading. Waste glasses were fabricated and tested that contained 3 wt % noble metals and 2.5 wt % MoO_3, respectively.

- The measured PCT normalized elemental releases for the glasses were at least an order of magnitude below that of EA glass.
- These measured T_L temperatures (1017 and 1128° C for the MoO_3 and noble metal limited glasses, respectively) would support glass processing at a nominal 1250° C processing temperature.

REFERENCES

1. "Standard Test Methods for Determining Chemical Durability of Nuclear, Hazardous, and Mixed Waste Glasses and Multiphase Glass Ceramics: The Product Consistency Test (PCT)," ASTM International, West Conshohocken, PA, ASTM C-1285, 2002.
2. "Standard Test Methods for Determining the Liquidus Temperature (TL) of Waste Glasses and Simulated Waste Glasses," Pacific Northwest National Laboratory, GDL-LQT, Rev. 4, 2007.
3. D.K. Peeler, T.B. Edwards, I.A. Reamer, J.D. Vienna, D.E. Smith, M.J. Schweiger, B.J. Riley, and J.V. Crum, "Composition/Property Relationships for the Phase 1 Am-Cm Glass Variability Study," Savannah River Site, Aiken, SC, WSRC-TR-99-00055, 1999.
4. D.K. Peeler, T.B. Edwards, T.S. Rudisill, I.A. Reamer, J.D. Vienna, D.E. Smith, M.J. Schweiger, and B.J. Riley, "Composition/Property Relationships for the Phase 2 Am-Cm Glass Variability Study," Savannah River Site, Aiken, SC, WSRC-TR-99-00393, 1999.
5. B.J. Riley, J.D. Vienna, M.J. Schweiger, D.K. Peeler, and I.A. Reamer, "Liquidus Temperature of Rare-Earth-Alumino-Borosilicate Glasses for Treatment of Americium and Curium," Mater. Res. Soc. Symp. Proc., 608 677-82 (2000).
6. J.C. Marra, D.K. Peeler, and C.M. Jantzen, "Development of Alternative Glass Formulations for Vitrification of Excess Plutonium," Savannah River Site, Aiken, SC, WSRC-TR-2006-00031, 2006.

7. C.L. Crawford, J.C. Marra, and N.E. Bibler, "Glass Fabrication and Product Consistency Testing of Lanthanide Borosilicate Glass for Plutonium Disposition," J. Alloys Compd., 444-445 569-79 (2007).
8. C.M. Jantzen, N.E. Bibler, D.C. Beam, C.L. Crawford, and M.A. Pickett, "Characterization of the Defense Waste Processing Facility (DWPF) Environmental Assessment (EA) Glass Standard Reference Material (U)." Savannah River Site, Aiken, SC, WSRC-TR-92-346 Rev.1, 1993.

CASTABLE GLASS AND GLASS-CERAMICS FROM DC PLASMA TREATMENT OF AIR POLLUTION CONTROL RESIDUES

D. Amutha Rani[a,b], J. A. Roether[a,b], D.E. Deegan[c], C. R. Cheeseman[a*] and A. R. Boccaccini[b*]

[a]Department of Civil and Environmental Engineering, Imperial College London, London SW7 2AZ, UK
[b]Department of Materials, Imperial College London, London SW7 2BP, UK
[c]Tetronics Ltd., South Marston Business Park, Swindon, Wiltshire SN3 4DE, UK
Corresponding authors: c.cheeseman@imperial.ac.uk and a.boccaccini@imperial.ac.uk

ABSTRACT

Energy from waste (EfW) plants combusting municipal solid wastes generate air pollution control (APC) residues from their emission abatement systems. APC residues are classified as hazardous waste, because of their high alkalinity, content of volatile heavy metals, high levels of leachable soluble salts, particularly chlorides, and trace organics including dioxins and furans; they have an absolute entry in the European Waste Catalogue. Consequently, the safe and effective management of APC residues is a major issue in the UK as they require physico-chemical treatment before disposal or thermal treatment to enable reuse. DC plasma arc technology is a recovery technology, producing a glass suitable for inert landfill but which has also been qualified for reuse as a product. In this work APC residues were combined with suitable fluxing agents and treated using DC plasma arc technology to produce a glass; this has been re-melted in the laboratory and cast into tiles. The tiles had comparable physical properties to commercial porcelain, granite and marble, with high bulk density (2900 kg/m^3) and hardness (6.2 GPa). The as-cast glass tiles were converted into glass-ceramics by treatment at 1000 °C. The crystalline phases formed were gehlinite ($Ca_2Al_2SiO_2$), anorthite ($CaAl_2Si_2O_8$) and magnesium silicate ($CaMgSi_2O_6$) and had an elongated, acicular morphology. The waste derived glass-ceramic tiles exhibited mechanical properties appropriate for use as heavy duty floor tiles.

INTRODUCTION

APC residues are regulated in the UK by the Environment Agency as hazardous waste, with an absolute entry in the European Waste Catalogue (19 01 07*)[1]. This is primarily because of their high alkalinity (pH >12) due to use of excess lime, and their content of volatile heavy metals, high levels of soluble chlorides and sulphate salts, and organic contaminants including dioxins and furans[2-4]. The total amount of APC residues produced in the UK is about 180,000 tonnes per annum, but this is expected to increase as the amount of municipal solid waste treated by incineration in EfW plants increases. An additional 40,000 tonnes per annum of APC residues will be generated in the UK from new EfW capacity in the near future, resulting from the impacts of the diversion targets associated with the landfill directive.

A number of technologies are available that may provide sustainable APC residue treatment[5]. Thermal treatment processes have previously been used to vitrify incinerator bottom ash[6,7] and incinerator fly ash[8], and APC residues from a major EfW plant in London (UK) have been processed using DC plasma arc technology[9]. The glass derived from plasma treatment has the potential for reuse in high grade applications such as paver sand, sand blasting, decorative cement and polymer-bound composites, but these are unlikely to be economically attractive as they would replace recycled glass. The objective of this work was to develop castable products directly from plasma-treated APC residues. This approach is attractive as it provides a direct way of transforming molten APC residues into useful products.

MATERIALS AND METHODS

APC residues were obtained from a major UK EfW plant. Crystalline phases present in APC residues are calcium hydroxide, calcium carbonate, and calcium hydroxychloride (CaCl(OH)) and the soluble salts sodium chloride and potassium chloride.

The mix fed into the DC plasma furnace consisted of 70 wt% APC residues with 21.9 wt% silica (99% sillimanite <1 mm grain size) and 8.3 wt% alumina (Chinese bauxite, 3-5 mm grain size)[8]. This mix was heated to approximately 1400 °C to form a homogeneous melt which cooled to form an apparently vitreous solid. XRD analysis of the product confirmed that it had an amorphous/glassy structure.

The APC residue derived glass (APCDG) was crushed to about 5 mm particle size and re-melted in the laboratory. The major oxide composition (Al_2O_3: 21 wt%, SiO_2: 38 wt% and CaO: 34 wt %), determined using XRF analysis[9], was used to calculate the change of melt viscosity with temperature, as shown in Figure 1[10].

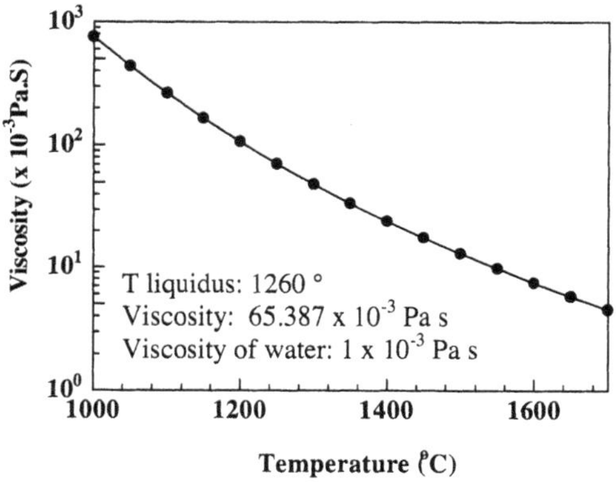

Figure 1. Change of viscosity with temperature for molten ARPDG obtained using the NPL method[10].

Temperatures of 1500 and 1600 °C were chosen in order to obtain a low viscosity, molten glass that would spread uniformly. The glass was cast into cast-iron moulds pre-heated to about 300 °C to minimise thermal shock effects. The bulk density of the as-cast glass samples was determined by Archimedes's method. Hardness was determined by Vickers indentation (Zwick/Roell Indetec ZHV) using a 500 g load applied for 10 seconds.

The development of glass-ceramics was also investigated. Glass tiles were heat treated for one and five hours at 850 °C and 1000 °C in order to produce glass-ceramic tiles. X-ray diffraction using CuKα radiation (XRD, Philips PW 1700) was used to determine the phases present in the heat-treated tiles and the microstructure of gold-coated polished surfaces was investigated by scanning electron microscopy (SEM, JEOL-JSM-840A).

RESULTS AND DISCUSSION

Figure 2 shows typical cast glass tiles made from APCDG. Rapid cooling of the melt resulted in a black glass consisting primarily of calcium, silicon and aluminium compounds. An XRD spectrum of the cast tiles is shown in Figure 3 and this confirms they are amorphous, with a broad background peak at around 30° (2θ) as for the original APCDG[9]. Chemical composition of the as-cast glass and ARPDG determined by XRF analysis indicated identical compositions.

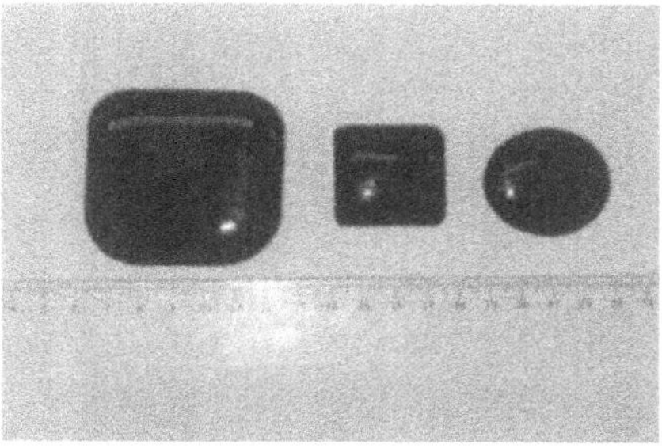

Figure 2. Cast glass tiles derived from re-melting ARPDG.

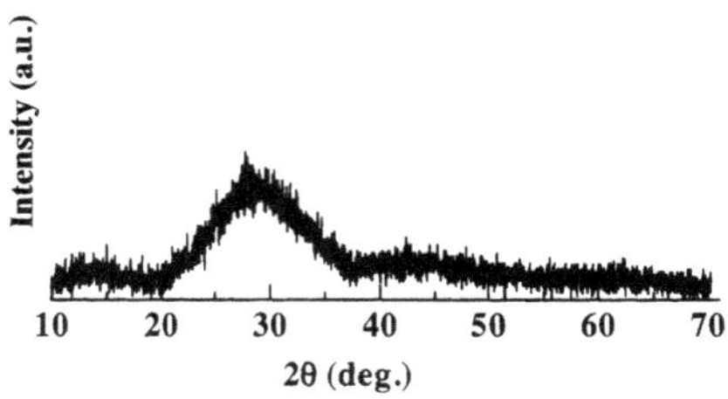

Figure 3. X-ray diffractogram of as-cast glass.

Material Property	Cast APC glass	Sintered APC glass-ceramics[13]	Granite [11,12]	Marble [11,12]	Porcelain stoneware tiles	Machineable glass-ceramics[13]
Density (kg/m^3)	2930	2400	2600- 2800	2600 - 2700	2400	2520
Hardness (Hv) GPa	6.2 ± 0.2	5.5 ± 0.1	5.1- 6.9	1.3 - 1.7	5.8	2.3

Table 1. Properties of cast glass and commercial products

Table 1 shows the material properties of as-cast glass tiles and sintered APC residue glass-ceramics, granite, marble, porcelain stoneware tiles and machineable glass-ceramics[11-13]. The bulk density (2900 kg/m^3) and hardness (6.2 GPa) values of as-cast tiles are high compared to those of commercial tiles. XRD analysis indicated that after heat-treating as-cast glass samples for one and five hours at 850 °C they remained amorphous. On heat-treating at 1000 °C the crystalline phases gehlinite ($Ca_2Al_2SiO_2$) anorthite ($Ca(Al_2Si_2O_8)$) and magnesium silicate ($CaMgSi_2O_6$) were formed.

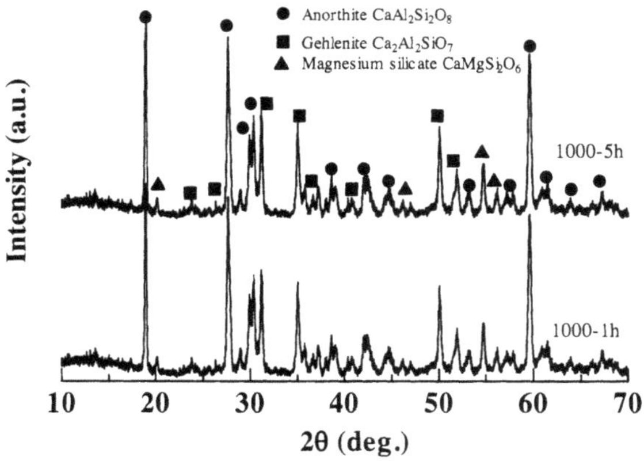

Figure 4. Phase analysis of heat treated cast glass.

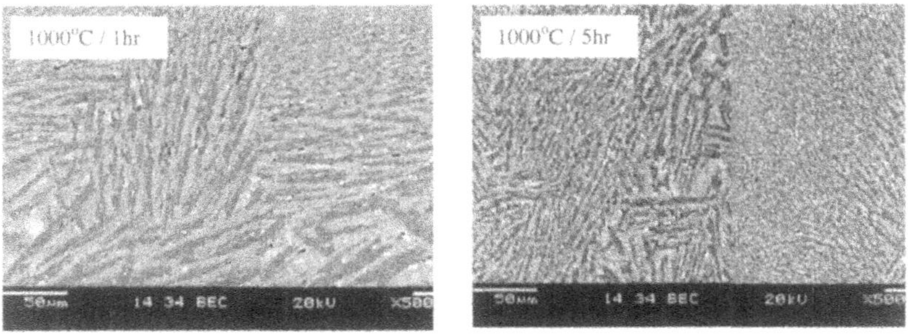

Figure 5. SEM image showing the microstructure of as-cast glass heat-treated at 1000 °C for different times.

Figure 5 shows the microstructures of tiles heat treated at 1000 °C for one and five hours to consist of a mainly acicular structure which could act as a crack-stopper and so contribute to increased toughness by crack deflection mechanism[15]. A cross section of the heat treated sample revealed that the elongated crystalline needle like phase was observed only on the surface and near pores (Fig. 6). This suggests that free surfaces are nucleating the crystalline phase due to superficial condensed phase impurities. The rate of diffusion will be much higher at a surface than in the bulk. Another influence could be chemical changes as a result of contact with atmosphere. Heterogeneous nucleation of crystalline phases is commonly observed at free surfaces as a result of chemical impurity concentrations and the greater rate of surface diffusion than bulk diffusion.

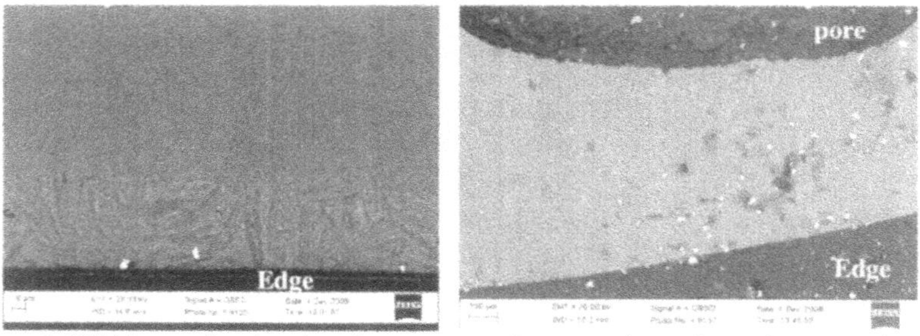

Figure 6. SEM image showing the microstructure of cross sections of cast glass-ceramics

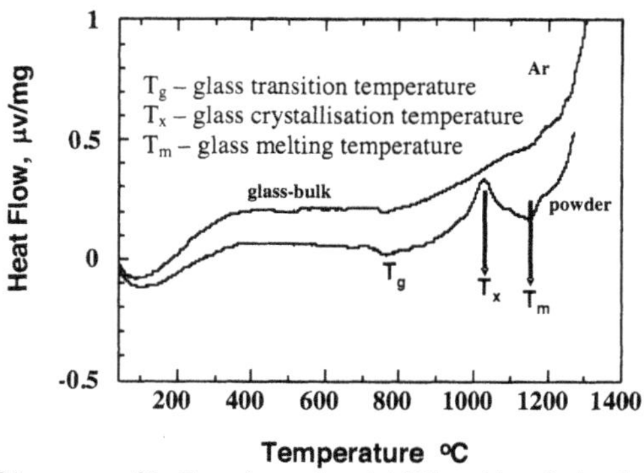

Figure 7. DTA curves of bulk and powdered APC residue derived glass.

The crystallisation behaviour of the APC derived glass was studied by differential thermal analysis (DTA). Results are shown in Figure 7 and the shallow and broad endothermic peaks indicate that surface crystallisation is the predominant mechanism for devitrification of the glass as is evident from the response of the powdered samples. The T_g/T_m ratio was 0.65 and this confirms heterogeneous nucleation and surface crystallisation[9]. While further investigation is required to establish the optimum heat treatment condition to develop glass-ceramics with optimum mechanical properties, these preliminary results indicate that ARPDG can be transformed into a glass-ceramic.

CONCLUSIONS

APC residues treated using DC plasma arc technology produces an inert glass. Glass tiles were formed by melting and casting the APC residues derived glass. These had high bulk density (2900 kg/m^3) and high hardness (6.2 GPa) compared to marble and granite tiles. Glass-ceramic tiles could also be produced directly from the DC plasma furnace by controlled cooling to allow the beneficial reuse of APC residues. The manufacture of products from the glass would further enhance the commercial performance and clear viability of the APC residue treatment using DC plasma technology.

ACKNOWLEDGEMENTS

This work was completed as part of the project 'Integrated solution for air pollution control residues (APC) using DC plasma technology' funded by the UK Technology Strategy Board and Defra, through the Business Resource Efficiency and Waste (BREW) programme.

REFERENCES

1. Environment Agency, Living Guidance from the Environment Agency, Using the list of wastes to code waste, 2006 Available at http://www.opdems.ac.uk/_files/LOW_Guide_v1%202%20(sustainability).pdf
2. DEFRA. Environmental Services Association (ESA), Draft APC residue case study. Available at http://www.defra.gov.uk/environment/waste/hazforum/pdf/hwf-3- 5iv.pdf (2005).
3. Environmental Services Association (ESA), Legislation and Policy. Available at http://www.esauk.org/waste/legislation/ (2005).
4. WRc Ref:/11873-0, Characterisation of Air Pollution Control residues from MSW Energy from waste. Resource Recovery Forum, (2004).
5. D. Amutha Rani, A. R. Boccaccini, D. Deegan, and C. R. Cheeseman, Air Pollution Control Residues from Waste Incineration: Current UK Situation and Assessment of Alternative Technologies, *Waste Manag.*, **28** 2279–2292 (2008).
6. S. Monteiro da Rocha., C. R. Cheeseman, S. Betanis, C. Sollars and A. R. Boccaccini, Ceramic Processing of Incinerator Bottom Ash, *Waste Manag.*, **23**, 907-916 (2003).
7. L. Barbieri, A. Corradi, I. Lancellotti, Bulk and Sintered Glass Ceramics by Recycling Municipal Incinerator Bottom Ash, *J. Eur. Ceram. Soc.*, **20**, 1637-1643 (2000).
8. A. R. Boccaccini, J. Schawohl, H. Kern, B. Schunck, J. M. Rincon, M. Romero, Sintered Glass Ceramics from Fly Ash, *Glass Technology*, **41**, 99-105 (2000).
9. D. Amutha Rani, E. Gomez, A. R. Boccaccini, L. Hao, D. Deegan, C. R. Cheeseman, Plasma Treatment of Air Pollution Control Residues, *Waste Manag.*, **28**, 1254-1262 (2008).
10. K. C. Mills and S. Sridhar, Viscosities of Ironmaking and Steelmaking Slags, *Ironmaking and Steelmaking*, ISSN 0301–9233, **26** 262- 269 (1999).
11. A. Karamanov, M. Pelino, M. Salvo, I. Metekovits, Sintered Glass-Ceramics from Incinerator Fly Ashes. Part II. The Influence of the Particle Size and Heat-treatment on the Properties, *J. Eur. Ceram. Soc.* **23**, 1609-1615 (2003).
12. A. Karamanov, M. Pelino, A. Hreglich, , Sintered glass-ceramics from Municipal Solid Waste-Incinerator Fly Ashes—Part I: The Influence of the Heating Rate on the Sinter-Crystallisation, *J. Eur. Ceram. Soc.*, **23,** 827-832 (2003).

13. Corning Product Information: http://www.corning.com/docs/specialty materials/pisheets/Macor.pdf. http://www.corning.com/WorkArea/showcontent.aspx?id=14893
14. D. Amutha Rani, A. R. Boccaccini, D. Deegan, and C. R. Cheeseman, Glass-Ceramic Tiles Prepared by Pressing and Sintering DC Plasma-Vitrified Air Pollution Control Residues, *Int. J. Appl. Ceram. Technol.*, **1–10** (2009) (in press)
15. M. Kotoul, J. Pokluda, P. Sandera, I. Dlouhy, Z. Chlup, and A.R. Boccaccini, Toughening Effects Quantification in Glass Matrix Composite Reinforced by Alumina Platelets, *Acta Materialia*, **56**, 2908-2918 (2008)
16. E. D. Zanotto, E. Muller, A Simple Method to Predict the Nucleation Mechanism in Glass, *J. Non-cryst. Solids,* **130**, 220–221 (1991).

XAFS OF Pu L_{III} EDGE IN LaBS GLASS

S.V. Stefanovsky, A.G. Ptashkin
SIA Radon
Moscow, RUSSIA

A.A. Shiryaev
Institute of Crystallography RAS
Moscow, RUSSIA

J.V. Zubavitchus, A.A. Veligjanin
RRC "Kurchatov Institute"
Moscow, RUSSIA

J.C. Marra
Savannah River National Laboratory
Aiken, SC, USA

M.V. Chukalina
Institute of Microelectronics Technology RAS
Chernogolovka, Moscow District, Russia

ABSTRACT

Plutonium-bearing LaBS glass with a target chemical composition (wt %): 9.0 Al_2O_3, 11.8 B_2O_3, 12.2 Gd_2O_3, 6.3 HfO_2, 17.2 La_2O_3, 13.6 Nd_2O_3, 9.5 PuO_2, 18.1 SiO_2, 2.3 SrO was prepared from PuO_2 powder that was mechanically mixed with chemicals. The batch was heated to a temperature of 1500 °C at a rate of 10 °C/min followed by holding for 30 min and cooling down to room temperature at a rate of 10 °C/min. The product was re-milled, heated and cooled by the same procedure in an attempt to better homogenize the glass. The product obtained was visually homogeneous. X-ray diffraction of the as-prepared glass indicated a major vitreous phase and minor concentrations of britholite and PuO_2. The fraction of crystalline constituent appeared to increase with storage time. Fourier and wavelet transforms of EXAFS spectrum of Pu L_{III} edge showed that the intensity of the peak due to first co-ordination sphere was much higher than that due to the second co-ordination sphere. This indicated that despite the fact that some Pu entered crystalline phases (PuO_2, britholite), a significant fraction of Pu remained in the vitreous phase. Plutonium in the glass is predominantly tetravalent. The oxygen environment of the Pu^{4+} ions in the vitreous phase resembles axially squeezed tetragonal pyramid with a coordination number ~5.

INTRODUCTION

Lanthanide-Borosilicate (LaBS) glass capable of dissolving up to ~10 wt.% PuO_2 was designed for immobilization of plutonium-bearing wastes,[1] whereas maximum PuO_2 concentration in conventional borosilicate nuclear waste glasses was found to be about 2 wt.%.[2] Although the LaBS glass has higher elemental leach rates than titanate ceramics[3] and may also devitrify with time depending on the thermal history it remains one of the most promising Pu-waste forms due to the ability to fabricate the waste form using routine glass melting technologies and the relative tolerance of the LaBS glass to impurities that may be associated with the Pu-waste. However, the information on its structure and Pu speciation is very limited and mostly based on suggestions and data on Th and Ce-bearing analogues.[4-6] Therefore, it is very important to investigate the structure of actual Pu-bearing LaBS glass. A recent report[7] presents X-ray diffraction (XRD) and scanning electron microscopy with energy dispersive spectral (SEM/EDS) data. The as-prepared glasses contained undissolved PuO_2 and when thermally soaked at 1250 °C a PuO_2-HfO_2 solid solution phase formed. In our previous works we have investigated the as-prepared glass with the same chemical composition and found that it contained minor britholite and that the Pu in the glass is predominantly tetravalent.[8-10] The present work extends our investigations on determination of local environment of Pu in the LaBS glass stored for 1 and 1.5 years.

EXPERIMENTAL

The Pu-bearing glass was prepared as follows. Plutonium nitrate solution in amount of 2 mL was diluted to 5 mL followed by interaction with 2M NaOH to produce $Pu(OH)_4$. The $Pu(OH)_4$ in amount of 89.5 mg was ignited to obtain PuO_2 (80.2 mg). The ^{239}Pu specimen contains 0.2 wt.% ^{241}Am (10% of total activity).

The milled PuO_2 was intermixed with the mixture of chemicals in an amount of 806.6 mg to obtain the glass composition (wt.%): 9.0 Al_2O_3, 11.8 B_2O_3, 12.2 Gd_2O_3, 6.3 HfO_2, 17.2 La_2O_3, 13.6 Nd_2O_3, 9.5 PuO_2, 18.1 SiO_2, 2.3 SrO. The mixture was placed in a Pt ampoule, heated in air to a temperature of 1500 °C at a rate of 10 °C/min, held for 30 min and cooled to room temperature at a rate of 10 °C/min. The obtained specimen consisted of glass and separate crystalline phase (probably Pu oxide). The specimen was re-milled and subjected to the thermal treatment identical to that described above. The final product was visually more homogeneous.

X-ray diffraction (XRD) measurements were performed using a DRON-4 diffractometer (Fe K_α radiation, λ = 1.936 Å). X-ray absorption fine structure (XAFS) spectra were recorded at the Structural Materials Science (STM) Beamline of the synchrotron source at RRC “Kurchatov Institute”. The glass samples were measured at room temperature either as dispersed powder or as pellets pressed from powder mixed with sucrose in the transmission mode using a Si(220) channel-cut monochromator and two air-filled ionization chambers. Fluorescence spectra were also acquired. Powders of chemically pure Pu oxides Pu_2O_3 and PuO_2 were used as standards and measured under identical conditions. Pu L_{III} edge in the LaBS glass was investigated after sample storage for 1 and for 1.5 years. Experimental XAFS spectra were fitted in R-space using an IFEFFIT package[11] and crystal structures of corresponding oxides and silicates. In the fitting, *ab initio* photoelectron backscattering amplitudes and phases calculated self-consistently using FEFF8[12] were used.

In the studied glass, Pu as well as other elements may enter several different phases and a XAFS signal obtained is a superposition of contributions from ions in different local environments. A similar situation arises when several different types of atoms are present at one distance from the core atom or when single and multiple scattering paths with identical path length exist. In such cases unambiguous analysis of the XAFS results using Fourier transform becomes problematic. To get a deeper insight into Pu behavior in LaBS glass we have applied wavelet transformation to experimental $\chi(k)$ curves.

Wavelet transform (WT) is commonly applied to evaluation of complex time-frequency signals. As shown in refs[13,14], WT is easily adapted to EXAFS analysis, and the expression of the WT of the k^n-weighted EXAFS data takes the form:

$$W_{\chi}^{\psi}(k,r) = (2r)^{1/2} \int_{-\infty}^{\infty} \chi(k')k'^{n} \psi^{*}[2r(k'-k)]dk', \quad (1)$$

where $\chi(k)$ is the EXAFS signal and $\psi^{*}[2r(k'-k)]$ is the complex wavelet function.

The WT is able to resolve the k dependence of the absorption signal, which potentially allows separation of contributing backscattering atoms even situated at the same distances from the core. One of the advantages of the wavelet analysis is the visualisation of the WT modulus in a k-R plot, which provides an easy way to interpret the results. Our analysis of EXAFS data for Pu and Hf were performed using the FORTRAN program HAMA employing Morlet wavelet algorithm.[15] The Morlet wavelet is well-suited for EXAFS signal since it consists of a slowly varying amplitude term and a fast oscillating phase term. Its mathematical description is broadly analogous to the Fourier transform. The Morlet wavelet is obtained by taking a complex sine wave with frequency η (as in FT) and by confining it with a Gaussian envelope with the half width σ,

$$\psi(k) = \frac{1}{(2\pi)^{1/2}\sigma} \exp(i\eta k)\exp(-k^{2}/2\sigma^{2}). \quad (2)$$

The choice of the η and σ parameters is important for data analysis since, besides other issues, it determines resolution in k-R space. Various combinations of these parameters were used in an attempt to resolve contributions from atoms at close distances from the central atom. As shown in ref.[14] use of higher k-weighting decreases resolution in the k-space, since backscattering amplitudes become flattened and shifted to higher values. Nevertheless, WT modulus plots for different k weights are shown to emphasize contributions of light and heavy backscatterers. Note, that in all plots of the WT modulus the interatomic distances are given without phase shift correction.

RESULTS AND DISCUSSION

X-ray diffraction of the as-prepared glass indicated a major vitreous phase and minor concentrations of PuO_2 (or PuO_2-HfO_2 solid solution with minor HfO_2), britholite and an oxide with a fluorite structure and a composition close to $GdHfO_{1.75}$. The fraction of crystalline constituent increased after glass storage at room temperature (stored between layers of sticky tape) for ~1.5 yrs (Figure 1). The glass with the same chemical composition studied in work[7] contained only minor PuO_2. Anyway phase composition depends strongly on preparation procedure and results are poorly reproducible.

Computer simulation of experimental XANES range of spectrum of Pu L_{III} edge in the glass studied (Figure 2) provided the best fit when ~90% of total Pu existed as Pu(IV) and ~10% was present as Pu(III). This result was in a good agreement with reference data[2,16,17] and our XPS data,[8,9] where almost all Pu was found to be in a tetravalent state.

Plots of FT magnitude of EXAFS range of spectra at Pu L_{III} edge after different storage periods (1 and 1.5 yr) are broadly similar but somewhat different from the curve for the as-prepared glass.[9,10] The peak due to first coordination sphere is much more intense than that of the second shell (Figure 3). This indicates that though some Pu enters crystalline phases (mainly as distorted PuO_2), its major fraction remains in the vitreous phase. However, XRD data shows that degree of crystallinity of the glass increases with time. This is also reflected in EXAFS spectra: the second and third coordination shells become more pronounced with increasing storage time. The best fits of the EXAFS spectra are given in Table I accounting for the phase shift. Oxygen environment of the Pu^{4+} ions in the vitreous phase resembles axially squeezed tetragonal pyramid with a coordination number ~5.

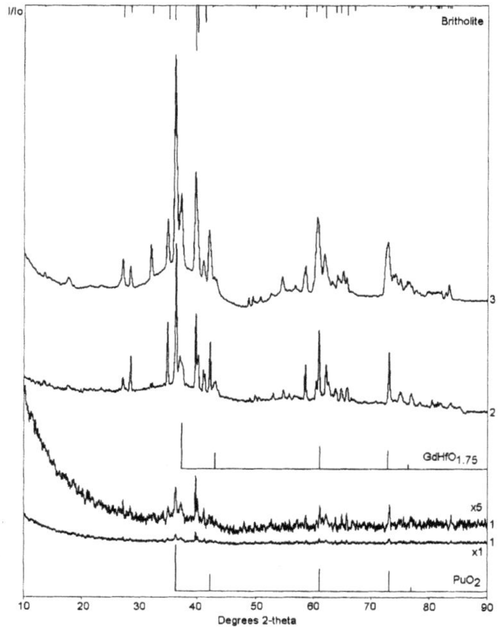

Figure 1. XRD patterns of as-prepared (1), stored for 1 yr (2) and 1.5 yr (3) LaBS Glass.

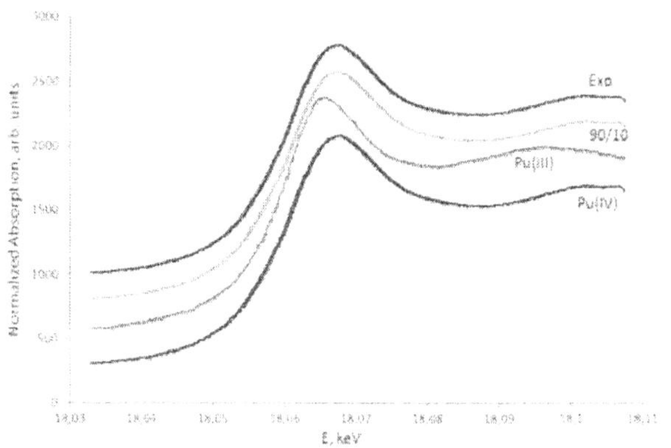

Figure 2. XANES spectra of Pu L_{III} Edge for Pu(III) in Pu_2O_3, Pu(IV) in PuO_2, in Pu-Bearing LaBS Glass (experimental curve) and the best fit at computer simulation at 90% Pu(IV) and 10% Pu(III).

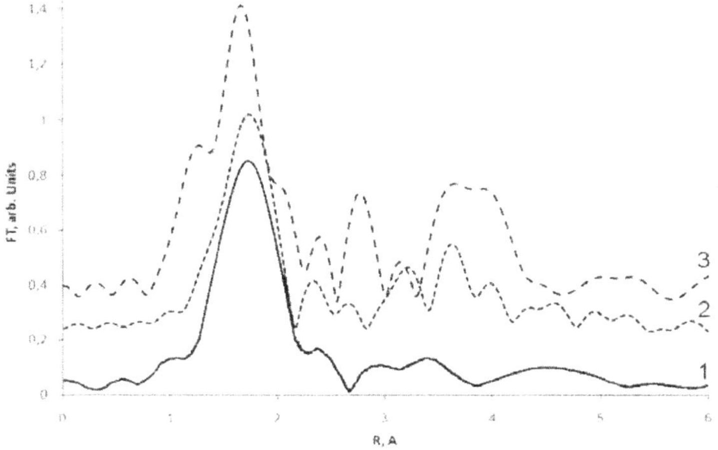

Figure 3. FTs of the EXAFS spectra of Pu L_{III} Edge in the as-prepared (1) and stored for 1 (2) and 1.5 yr (3) LaBS Glass (the data are given without phase shift).

Table I. Fitting of the Pu L_{III} Edge EXAFS Spectra of the LaBS Glass.

Stored for 1 yr				Stored for 1.5 yr			
Bond	Population	Sigma	Distance, A	Bond	Population	Sigma	Distance, A
Pu—O_{1a}	0.03	0.03	1.91	Pu—O_{1a}	~1	0.04	1.98
Pu—O_{1b}	0.29	0.01	2.07	Pu—O_{1b}	4	0.04	2.18±0.03
Pu—O_{1c}	5.54	0.005	2.28	Pu—O_{1c}	2.7	0.04	2.35±0.03
Pu—Pu(M)	3.58	0.005	3.73	Pu—Pu(M)	3.8	0.04	3.72
Pu—O_2	1.32	0.01	4.4	Pu—O_2	-	0.06	4.4

Closer look at the FT plots reveals that the peak due to the Pu-O bond in the first coordination sphere in the structure of the LaBS glass had significant asymmetry (Figure 3) due to overlapping of the peaks due to Pu-O bonds in the vitreous phase and crystalline PuO_2 in the glass. WT magnitude plots are shown on Figure 4. Analysis using Morlet wavelet suggests that a small amount of a heavy element was also present at ~2.2 Å. Whereas, at least partially, this signal is due to Pu in PuO_2 (well pronounced on Figure 3), contribution of other heavy glass constituents is plausible. Unfortunately, the separation of oxygen and the heavy element in k-R space is insufficient for precise determination of the k-coordinate for the heavy element even when the optimized wavelet is used. At a distance of 2 Å from the central atom the backscattering amplitudes for Pu, Hf and REE are very close between k values 7 and 10 $Å^{-1}$. Therefore, from the current EXAFS results unambiguous identification of the second (non-Pu) heavy element in the splitted first coordination sphere is impossible. Unambiguous identification of this element is barely possible due to strong overlap with O contributions, but Hf seems to be a plausible candidate. Since similar splitting of the first coordination sphere is observed around Hf ions, this FT peak could be tentatively assigned to contribution from PuO_2-HfO_2 solid solution.

The peak with maximum at ~3.72-3.73 Å (located at ~2.8 Å in spectra without phase shift (Figure 3)) is mostly due to Pu-Pu distance in PuO_2, though Pu-M (M = Si, La, Nd, Gd, Hf) pairs in the vitreous phase may also contribute. Occurrence of REE (La, Nd, Gd) or Hf at distances of ~5.2 Å is plausible, but is hard to prove conclusively due to low statistics at high k-values.

Wavelet analysis allows evaluation of glass structure evolution with time. The maxima corresponding to different elements on the WT plots with various k-weighting become sharper with storage, strongly supporting idea that the glass devitrifies and Pu dioxide and britholite grains are precipitated, the grain size and crystallographic perfection increases. In the glass stored for 1 yr, atoms at the (uncorrected) distance around 3 Å are represented mostly by heavy elements (Pu and/or REE), and the oxygen fraction is minor. In contrast, in the glass stored for 1.5 yr this shell is dominated by O, probably indicating diffusion of heavy elements towards crystalline grains. Sharpening of the maxima in R-space indicates better separation of contributions from glassy and crystalline phases. Comparison of the k^3-weighted WT plots supports assignment of the element at 2.2 Å to Pu, but contribution of different atoms is also clear from asymmetric shape of the maximum.

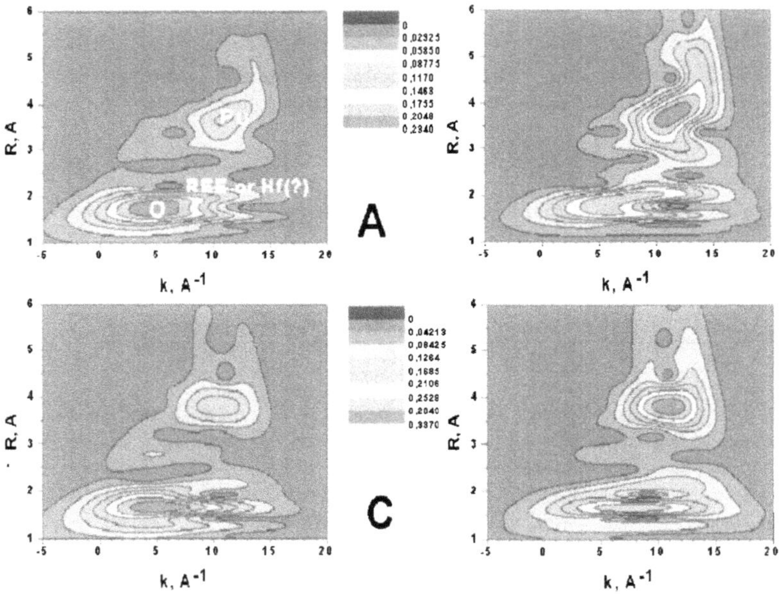

Figure 4. WT magnitude plots for Pu in the LaBS glass with k-weighting of 2 (A, C) and 3 (B, D) stored for 1 yr (A, B) and 1.5 yrs (C, D).

CONCLUSION

The as-prepared LaBS glass containing 9.5 wt.% PuO_2 is composed of a predominant vitreous phase and minor PuO_2, britholite-type phase and an oxide with a fluorite structure and a composition close to $GdHfO_{1.75}$. The glass appeared to devitrify to some degree with time. The fraction of crystalline constituent increased after glass storage for ~1.5 yrs. Plutonium in the glass is predominantly tetravalent. Oxygen environment of the Pu^{4+} ions in the vitreous phase resembles an axially squeezed tetragonal pyramid with a coordination number ~5. For the sample exposed for ~ 1.5 yrs the fit of the spectrum (R-factor = 0.02) gives the following distances: R $(Pu\text{-}O_1)_1$ = 1.98 Å (σ = 0.04), R $(Pu\text{-}O_1)_2$ = 2.18±0.01 Å (σ = 0.04), R $(Pu\text{-}O_1)_3$ = 2.355±0.015 Å, R (Pu-Pu) = 3.72 Å (σ = 0.04), R $(Pu\text{-}O_2)$ = 4.4 Å (σ = 0.06).

REFERENCES

[1] N.E. Bibler, W.G. Ramsey, T.F. Meaker and J.M. Pareizs, Durabilities and Microstructures of Radioactive Glasses for Immobilization of Excess Actinides at the Savannah River Site, *Mater. Res. Soc. Symp. Proc.* **412**, 65-70 (1996).
[2] B.W. Veal, J.N. Mundy, and D.J. Lam. Actinides in Silicate Glasses, *Handbook of the Physics and Chemistry of Actinides*, A.J. Freeman and G.H. Lander (eds.). Elsevier Science Publishers B.V., 1987. P. 271-309.
[3] D.M. Strachan, A.J. Bakel, E.C. Buck, D.B. Chamberlain, J.A. Fortner, C.J. Mertz, S.F. Wolf, W.F. Bourcier, B.B. Ebbinghaus, H.F. Shaw, R.A. Van Konynenburg, B.P. McGrail, J.D. Vienna, J.C. Marra and D.K. Peeler. The Characterization and Testing of Candidate Immobilization Forms for the Disposal of Plutonium, *Waste Management '98 Conf.* (Laser Options Inc., Tucson, AZ, 1998). ID 65-08, CD-ROM.
[4] M.G. Mesko, T.F. Meaker, W.G. Ramsey, J.C. Marra, D.K. Peeler. Optimization of Lanthanide Borosilicate Frit Compositions for the Immobilization of Actinides Using a Plackett-Burman/Simplex Algorithm Design, *Mater. Res. Soc. Symp. Proc.* **465**, 105-110 (1997).
[5] T.F. Meaker, D.K. Peeler, J.C. Marra, J.M. Pareizs, and W.G. Ramsey. Actinide Solubility in Lanthanide Borosilicate Glass for Possible Immobilization and Disposition, *Mater. Res. Soc. Symp. Proc.* **465**, 1281-1286 (1997).
[6] J.C. Marra, C.L. Crawford, and N.E. Bibler. Glass Fabrication and Product Consistency Testing of Lanthanide Borosilicate Frit B Composition for Plutonium Disposition, WSRC-TR-2006-00033, January 2006, Washington Savannah River Company, Aiken, SC (2006).
[7] J.C. Marra, C.L. Crawford, and N.E. Bibler. Glass Fabrication and Product Consistency Testing of Lanthanide Borosilicate Frit X Composition for Plutonium Disposition. WSRC-STI-2006-00318. SRNL (2006).
[8] K.I. Maslakov, S.V. Stefanovsky, A.Yu. Teterin, Yu.A. Teterin, and J.C. Marra. X-Ray Photoelectron Study of Lanthanide-Borosilicate Glass, *Glass Phys. Chem.* **35**, 22-28 (2008).
[9] J.C. Marra, S.V. Stefanovsky, A.A. Shiryaev, Y.V. Zubavichus, and A.G. Ptashkin. Preparation and Characterization of the Pu-Bearing LaBS Glass. *Plutonium Future – "The Science" 2008. A Topical Conference on Plutonium and the Actinides*. Dijon, France, 7-11 July, 2008. Abstracts. 102-103 (2008).
[10] S. Stefanovsky, A. Shiryaev, Y. Zubavichus, J. Marra. Plutonium Environment in Lanthanide-Borosilicate Glass. *39^{iemes} Journees des Actinides*, March 28-31, 2009. La Grande Motte, France. Abstracts. 123-124 (2009).
[11] B. Ravel and M. Newville. ATHENA, ARTEMIS, HEPHAESTUS: Data Analysis for X-ray Absorption Spectroscopy Using IFEFFIT, *J. Synchrotron Radiat.*, **12**, 537-541 (2005).
[12] A.L. Ankudinov and J.J. Rehr. Relativistic Calculations of Spin-Dependent X-ray-Absorption Spectra, *Phys. Rev.* B **56**, 1712-1716 (1997).
[13] H. Funke, A.C. Scheinost, M. Chukalina. Wavelet Analysis of Extended X-ray Absorption Fine Structure Data, *Phys. Rev. B*, **71**, 094110 (2005).

[14] H. Funke, M. Chukalina, A.C. Scheinost, A New FEFF-Based Wavelet for EXAFS Data Analysis, *J. Synchrotron Radiat.* **14**, 426-432 (2007).
[15] www.esrf.fr/exp_facilities/BM20/Software/Wavelets.html
[16] P.G. Eller, G.D. Jarvinen, J.D. Purson, R.A. Penneman, R.R. Ryan, F.W. Lytle, and R.B. Greegor, Actinide Valences in Borosilicate Glass, *Radiochim. Acta.* **39**, 17-22 (1985).
[17] N.J. Hess, W.J. Weber, S.D. Conradson X-ray absorption fine structure of aged, Pu-doped glass and ceramic waste forms, *Mat. Res. Soc. Symp. Proc.* **506**, 169-176 (1998).

PHASE FORMATION STUDIES USING X-RAY DIFFRACTION AND INFRARED SPECTROSCOPY IN THE VITRIFICATION OF SAVANNAH RIVER SITE SB4 HLW SLUDGE SURROGATE WITH HIGH IRON AND ALUMINUM CONTENTS AT HIGH WASTE LOADINGS

O.I. Stefanovsky, S.V. Stefanovsky, D.Y. Suntsov
SIA Radon
Moscow, RUSSIA

A.A. Akatov
Institute of Technology
St-Petersburg, RUSSIA

J.C. Marra
Savannah River National Laboratory
Aiken, SC, USA

ABSTRACT

Phase formation mechanisms associated with the vitrification of Savannah River Site (SRS) Sludge Batch 4 (SB4) high level waste surrogate with high iron and aluminum contents were studied by X-ray diffraction (XRD) and infrared spectroscopy (IRS). Two mixtures at 60 and 70 wt% SB4 waste loading were prepared as slurries with a water content of ~50 wt% using a waste surrogate and commercially available Frit 503-R4 (Li_2O – 8 wt %, B_2O_3 – 16 wt %, SiO_2 – 76 wt %). The mixtures were air-dried at a temperature of 115 °C and heat-treated at 500, 700, 900, 1000, 1100, 1200, and 1300 °C for 1 hr at each temperature. IR spectra and XRD patterns of the products heat-treated at each temperature were recorded. In both the mixtures phase formation reactions started at low temperatures and yielded intermediate phases (sodalite, pyroxene-type, nepheline), and reactions were mostly completed within the temperature range between 1000 and 1100 °C. The glassy materials prepared at 1200 °C were composed of vitreous phase and magnetite/trevorite type spinel. The materials with 60 wt.% waste loading prepared at 1300 and 1400 °C had the same phase composition whereas at 70 wt.% waste loading, nepheline phase was also detected resulting from crystallization from the melt.

INTRODUCTION

Vitrification is currently considered as one of the most promising method of high level waste (HLW) conditioning in USA, Russia, France, UK, Germany, Japan, Korea and other countries.[1] In the USA a Joule heated ceramic melter (JHCM) is used at the Defense Waste Processing Facility (DWPF) for treatment of liquid HLW stored in tanks at Savannah River Site.[2] Due to process constraints maximum waste loading in glass is limited by ~40 wt.%.[3] Increase of waste loading in glass would allow for a higher volume reduction factor and corresponding cost savings. This may be reached by application of cold crucible inductive melting (CCIM) technology where the active hydrodynamic regime in the melter provides for higher waste capacity and glass productivity. Moreover, the melter of this type does not have refractories and electrodes contacting the melt that increases its lifetime and reliability.[4]

The CCIM provides the possibility for increased waste loadings and higher melting temperature compositions. An understanding of chemical reactions occurring during the glass melting process is necessary to understand the melting behavior of these glasses as well as the crystallization behavior (and the relationship to crystal formation in the melter). In our previous work[5] we investigated phase formation mechanism in the mixtures of Sludge Batch 2 (SB2) high-ferrous HLW and either commercially available Frit 320 or mixture of chemical simulating this frit and found that reactions in the mixture with actual frit were completed at temperatures by 100-200 °C lower than in the mixture with chemicals and the glasses obtained were more homogeneous. In the present work we studied phase formation reactions in two mixtures at 60 and 70 wt% Sludge Batch 4 (SB4) waste loading prepared as slurries with a water content of ~50 wt% using a waste surrogate and commercially available Frit 503-R4 (Li_2O – 8 wt %, B_2O_3 – 16 wt %, SiO_2 – 76 wt %).

EXPERIMENTAL

SB4 waste surrogate (Table I) was prepared according to SRT-MST-2007-00070 procedure,[6] including precipitation of MnO_2 and Fe, Ni hydroxides from nitrate solutions.

Table I. Target Chemical Compositions of SB4 Waste Surrogate, Frit 503-R4 and Glasses

Component	Calcined SB4 Waste	Frit 503-R4	Glasses at Waste Loading	
			60 wt.%	70 wt.%
Al_2O_3	27.56	-	16.08	18.76
BaO	0.08	-	0.05	0.05
CaO	3.00	-	1.78	2.07
Cr_2O_3	0.22	-	0.13	0.15
CuO	0.05	-	0.03	0.04
Fe_2O_3	31.34	-	18.26	21.30
K_2O	0.08	-	0.05	0.06
MgO	3.00	-	1.79	2.09
MnO	6.25	-	3.63	4.24
Na_2O	20.22	-	12.71	14.83
NiO	1.80	-	1.04	1.22
PbO	0.41	-	0.21	0.25
SiO_2	2.93	76.00	32.13	24.79
ZnO	0.05	-	0.03	0.04
ZrO_2	0.10	-	0.06	0.07
F	0.02	-	0.01	0.01
Cl	1.59	-	0.97	1.14
I	0.04	-	0.02	0.03
P_2O_5	0.32	-	0.20	0.23
SO_3	0.94	-	1.22	1.43
B_2O_3	-	16.00	6.40	4.80
Li_2O	-	8.00	3.20	2.40
Total	100.00	100.00	100.00	100.00
Calcination factor	0.73	-	0.82	0.79

Two mixtures at 60 wt% and 70 wt.% SB4 waste loading with commercially available Frit 503-R4 (Li_2O – 8 wt.%, B_2O_3 – 16 wt.%, SiO_2 – 76 wt.%) labeled as 60/40 and 70/30 were prepared as slurries with a water content of ~50 wt% (Table I). The mixtures were air-dried at a temperature of 115 °C and heat-treated in a resistive furnace at 500, 700, 900, 1000, 1100, 1200, and 1300 °C for 1 hr at each temperature followed by removal of crucibles from the furnace, cooling in air and sampling of specimens for analyses. Portions of melts prepared at 1200, 1300, and 1400 °C were poured onto a metal plate and quenched in air. The phase formation mechanism was studied by X-ray diffraction (XRD) using a DRON-4 diffractometer (Fe K_{α} –radiation, λ = 1.936 Å, scan range of 10 to 90 degrees 2-theta) and infrared spectroscopy (IRS) using a IKS-29 spectrophotometer (compaction of powdered glass in pellets with KBr).

RESULTS AND DISCUSSION

XRD patterns of the products of heat-treatment of the waste/frit mixtures at 60 wt.% waste loading are shown on Figure 1. The d-spacings of the phases were also calculated to facilitate identification of the phases at each heat treatment temperature. In the mixture dried at a temperature of 115 °C gibbsite, $Al(OH)_3$, sodium nitrate (nitratine, $NaNO_3$), sodium sulfate (tenardite, Na_2SO_4) are major phases and spinel, nepheline and, possibly, halite (NaCl) and hematite (Fe_2O_3) are minor phases.

Within the temperature range of 115 to 500 °C gibbsite and major fraction of nitratine have reacted yielding a pyroxene or sodalite-type phase whereas tenardite and spinel remained nearly unaltered. At higher temperature (700 °C) pyroxene-type phase, spinel and tenardite as well as an unidentified phase were present in the product. In the product heat-treated at a temperature of 900 °C only pyroxene-type phase and spinel were found (vitreous phase is present in all the samples because they contain glass frit). Same phase composition occurs in the product heat-treated at 1000 and 1100 °C but the fraction of spinel grows with temperature whereas amount of the pyroxene-type phase decreases.

The products of heat-treatment of the mixtures at higher temperatures (1200, 1300 and 1400 °C) are composed of major vitreous and minor spinel phase. Fraction of the spinel phase is higher in the sample slowly cooled (annealed) in the crucible as compared with the sample quenched onto a metal plate.

Infrared (IR) spectra (Figure 2) also demonstrate gradual changes with temperature. Within the high wavenumber range spectra of the products contain a broad band at 3500-3700 cm^{-1} due to stretching vibrations H—O—H in molecules of absorbed and structurally-bound water. This band in the spectrum of the product dried at 115 °C (*1*) is overlapped with narrow components due to stretching O—H vibrations in hydroxides. Weak narrow bands at ~2755 and ~2427 cm^{-1} may be attributed to vibration of hydrogen bonds. The band with a maximum at 1613-1624 cm^{-1} as well as some bands (1505 cm^{-1} and others), overlapped with a strong narrow band with a maximum at ~1380 cm^{-1} due to stretching vibrations of N—O bonds in NO_3^- and NO_2^- and C—O bonds on CO_3^{2-} ions, may be assigned to bending vibrations of H—O—H bonds. The bands within the range of 1200 to 900 cm^{-1} and centered at about 550 cm^{-1} are, respectively, due to stretching and bending vibrations in SiO_4 groups in the glassy Frit 503-R4. Narrow bands within the range between 700 and 900 cm^{-1} are due to bending vibrations in NO_3, NO_2, CO_3 groups.[7,8] Broad band at 550-700 cm^{-1} is a superposition of stretching vibrations of Fe—O, Al—O, Mn—O bonds in spinel and other compounds.[9] Increase in the heat-treatment temperature reduces the content of volatile anions although their traces are still present even in the products heat-treated at 1000 °C.

IR spectra of the glassy materials produced at temperatures 1100 °C and higher consist of strong broad bands at 900-1200 and 400-550 cm^{-1} and weaker bands at ~1370-1390, ~1270-1275, ~700 and ~555-565 cm^{-1}. The bands with maxima at 1395–1410 cm^{-1} and ~1270 cm^{-1} are components of doubly degenerate asymmetric stretching, ν_3 O–B–O vibration in triangular BO_3 units. The bands at ~720 cm^{-1} and ~655 cm^{-1} are components of doubly degenerate asymmetric bending, δ_4, O–B–O vibrations.[10] The broad band at 1200–850 cm^{-1} with a maximum at ~1000 cm^{-1} is due to overlapping of the bands arising from asymmetric stretching of bridging Si–O–Si and nonbridging O–Si–O bonds in the silicon–oxygen glass network. This band is normally a superposition of the bands due to O—

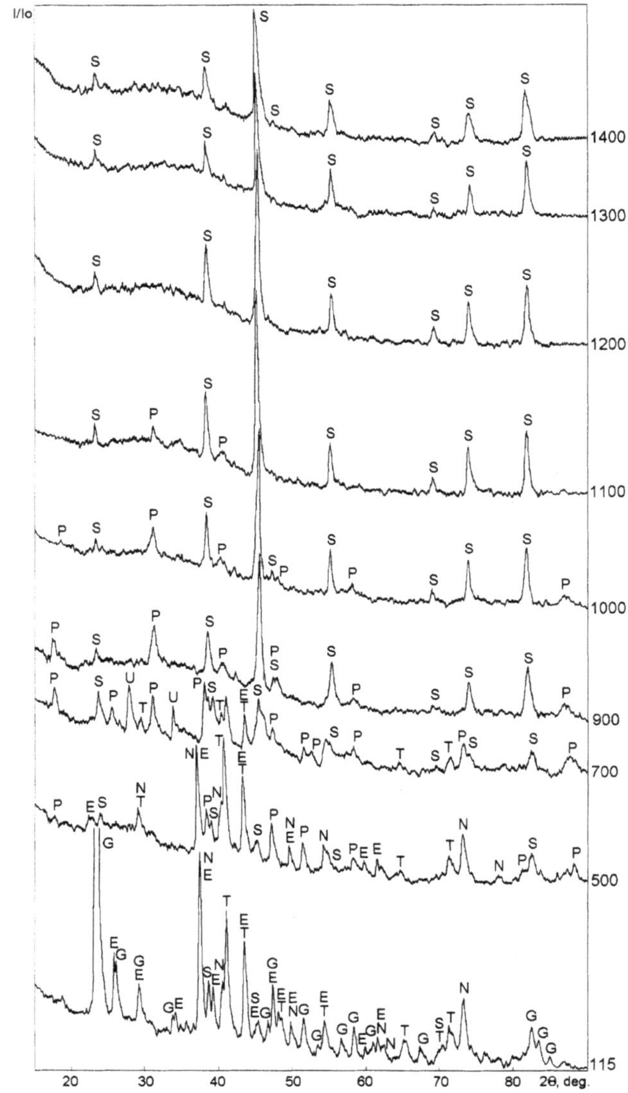

Figure 1. XRD Patterns of the Mixtures from SB4 Waste Surrogate (60 wt.%) and Frit 503-R4 (40 wt.%) Heat-Treated at Various Temperatures.
G – gibbsite, E – nepheline, N – nitratine, P – pyroxene- or sodalite-type phase, S – spinel, T – tenardite, U – unidentified.

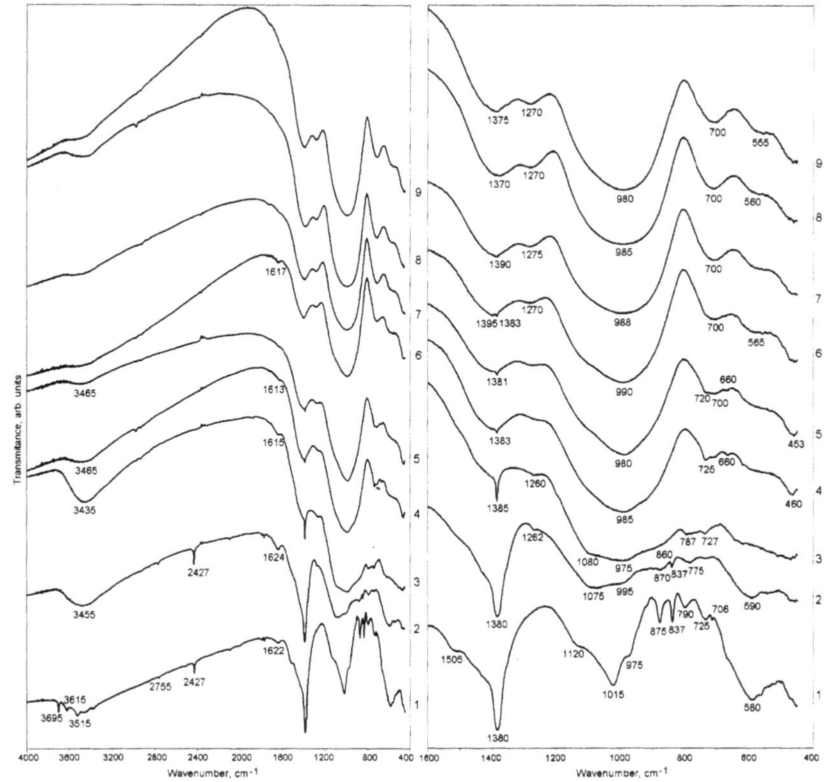

Figure 2. IR Spectra within the Range of 4000 to 400 cm^{-1} and Their Fragments of 1600 to 400 cm^{-1} of the Mixtures from SB4 Waste Surrogate (60 wt.%) and Frit 503-R4 (40 wt.%) Heat-Treated at Various Temperatures, °C: 1 – 115, 2 – 500, 3 – 700, 4 – 900, 5 – 1000, 6 – 1100, 7 – 1200, 8 – 1300, 9 – 1400.

Si—O vibrations in the SiO_4 units with various numbers of bridging oxygen ions connecting SiO_4 tetrahedra in the network: four (Q^4) – at ~1100-1150 cm^{-1}, three (Q^3) – at ~1050-1100 cm^{-1}, two (Q^2) – at ~1000-1050 cm^{-1}, one (Q^1) – at ~950-1000 cm^{-1}, and zero (Q^0) – at ~900-950 cm^{-1}. Maximum of the valence band is positioned at ~980-990 cm^{-1}. This means that the structure of anionic motif of the glass network is built from pyrosilicate and rather short metasilicate chains. The wavenumber range of 400-550 cm^{-1} corresponds to deformation vibrations in the SiO_4 units.[7-9] Boron exists as primarily BO_3 groups not incorporated in silica-oxygen chains but forming separate glass network constituent.

XRD patterns of the products of heat-treatment of the waste/frit mixtures at 70 wt.% waste loading are shown on Figure 3. The d-spacing values for each phase were calculated to assist in phase identification.

The mixture air-dried at 115 °C is composed of major nitratine, tenardite and spinel, and minor nepheline and halite (NaCl). Heat-treatment of the mixture at 500 °C resulted in a significantly lower amount of nitratine, which interacted with alumina and borosilicate frit yielding pyroxene or sodalite-type phase. At higher temperatures (700-900 °C) the mixtures are composed of glass, spinel and pyroxene or sodalite-type phase whereas tenardite decomposes. The borosilicate melt dissolves pyroxene-type phase within the temperature range between 1000 and 1100 °C and spinel remains the only crystalline phase in the products obtained at 1100, 1200 °C and in the products melted at 1300 and 1400 °C and quenched by pouring onto a metal plate. The same compositions heat treated at 1300 and 1400 °C in crucibles and cooled down to room temperature contained nepheline. This indicates that the nepheline is a product of crystallization from the melt rather than product of phase formation reactions.

IR spectra of the products of heat-treatment of the waste/frit mixtures are shown on Figure 4. They are similar to those of the mixtures with 60 wt.% waste loading. In the high wavenumber range the bands due to O—H and hydrogen bonds are present. Within the range between 1600 and 400 cm^{-1} spectrum of the air-dried sample consists of the strong bands due to stretching and bending vibrations of the bonds in nitrate, nitrite and carbonate groups (~1350-1400, 876, 837, 705 cm^{-1}), broad bands at 1200-900 cm^{-1} with maxima at 1120 and 1010 cm^{-1} and lower 550 cm^{-1} due to stretching and bending vibrations of Si—O bonds in SiO_4 tetrahedra, and 700-500 cm^{-1} due mainly to Fe—O and Al—O vibrations. The latter band is markedly stronger than that observed in the spectrum of the air-dried 60/40 sample higher waste loading. The components of 1120 and 1010 cm^{-1} may be assigned to stretching modes in the SiO_4 tetrahedra with various number of bridging oxygen ions in the structure of the glassy frit 503-R4: three, and two, respectively.

Intensities of the bands in the IR spectra due to hydroxyl groups and volatile anions decrease with temperature and disappear after heat-treatment of the 70/30 mixture at ~1000 °C (Figure 4). Behavior of the bands due to vibrations in the anionic motif of the glass structure is more complicated. The band 1200-800 cm^{-1} undergoes gradual transformation with a loss of the high-wavenumber component and symmetrization with formation of a nearly symmetric band with a maximum at ~975 cm^{-1}. This indicates a low degree of connectedness of the glass network which is mainly built from short pyrosilicate chains. In the whole, as it is seen from IRS, the structure of the material with 70 wt.% waste loading is significantly different from the structure of the glassy material with 60 wt.% waste loading. The spectra of the materials melted at 1200-1400 °C consist of the bands with maxima at 1373-1380, ~975, 715-717, ~460 cm^{-1} and two shoulders at ~1200-1210 and ~545 cm^{-1}.

The band centered in the vicinity of 1400 cm^{-1} is normally observed in silicate glasses but its attribution is contradictory. This band cannot be assigned to B—O—B stretching modes due to low B_2O_3 content (target value is 4.80 wt.%) in the material. There is an opinion that this band arises in IRS of glasses containing residual CO_3^{2-} ions.[11] Because this band is normally observed in glasses with high alkali and alkali earth contents[11,12] having low durability with respect to water and wet atmosphere, this point of view might be supported. Samples for IRS analysis are prepared from fine-grained powders which may absorb water and carbon dioxide from air yielding hydroxides and carbonates on their surface.

The band due to stretching modes of Si—O bonds in the SiO_4 tetrahedra consists of two well-resolved components in spectra of the products obtained at 1100 and 1200 °C, the resolution worsens in the spectrum of the material melted at 1300 °C and in the spectrum of the material melted at 1400 °C this band is nearly symmetric with a maximum at ~975 cm^{-1}. Because target chemical composition of the materials produced at various temperatures is the same, the changes observed may be accounted for incomplete melt homogenization even at temperatures of 1200-1300 °C due to high melt viscosity. As follows from XRD data, segregation of the spinel phase in the melt is also incomplete even at a temperature of 1300 °C (Figure 3). As a result, the structure of the melt is non-uniform and composed of network fragments with higher and lower degree of connectedness responsible for higher and lower wavenumber constituents of the band. Complete homogenization takes place at the highest melting temperature (1400 °C). At this temperature, the spinel phase is segregated from the vitreous phase (Figure 3).

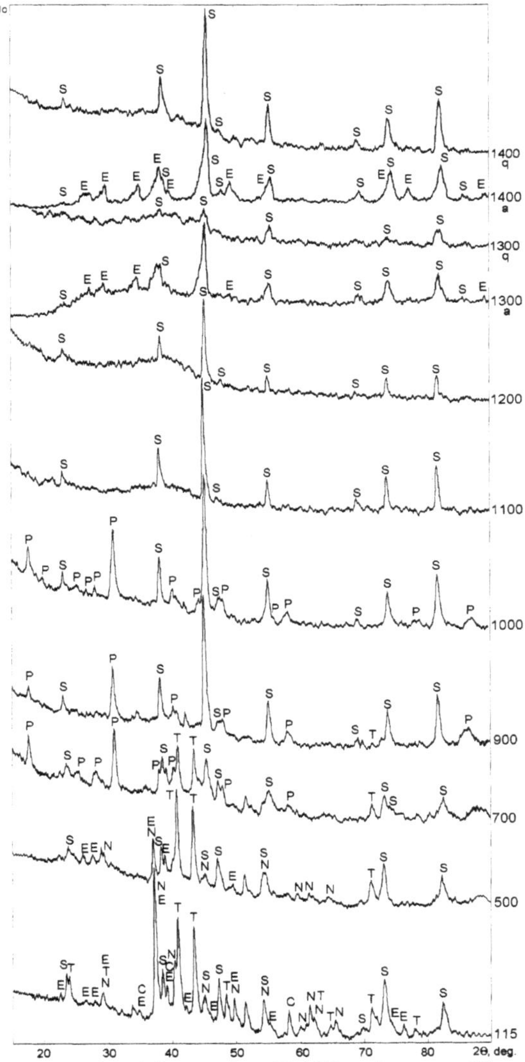

Figure 3. XRD Patterns of the Mixtures from SB4 Waste Surrogate (70 wt.%) and Frit 503-R4 (40 wt.%) Heat-Treated at Various Temperatures (°C).
C – halite, E – nepheline, N – nitratine, P – pyroxene-type phase, S – spinel, T – tenardite; a – annealing, q – quenching.

The band centered at 715-717 cm^{-1} arises from superposition of the bands due to stretching modes of Al—O bonds in the glass network and symmetric ν_1 stretching modes of Si—O bonds in the SiO_4 tetrahedra. Position of maximum of ν_3 Si—O$^-$ band at ~975 cm^{-1} points to predominant pyro-groups (SiO_4 tetrahedra with one bridging oxygen ion) in the glass structure. Therefore, these tetrahedra have symmetry lower than T_d. In this case the ν_1 symmetric vibrations become active in IR spectra and their absorption band would be positioned at 750-800 cm^{-1}.[7]

The shoulder at ~545 cm^{-1} may be attributed to superposition of the bands due to Fe—O vibrations in tetrahedral FeO_4 units in the spinel structure, Al—O vibrations in octahedral AlO_6 units in both the spinel and glass structures,[9] and bending ν_4 vibrations in the SiO_4 units.

Attribution of the weak band (shoulder) at ~1200-1210 cm^{-1} is debatable. Because this band is observed in IR spectra of boron-containing glasses, it may be assigned to vibrations in complex borate groups with three-coordinated boron. Low boron content in the 70/30 material explains low intensity of absorption in this range.

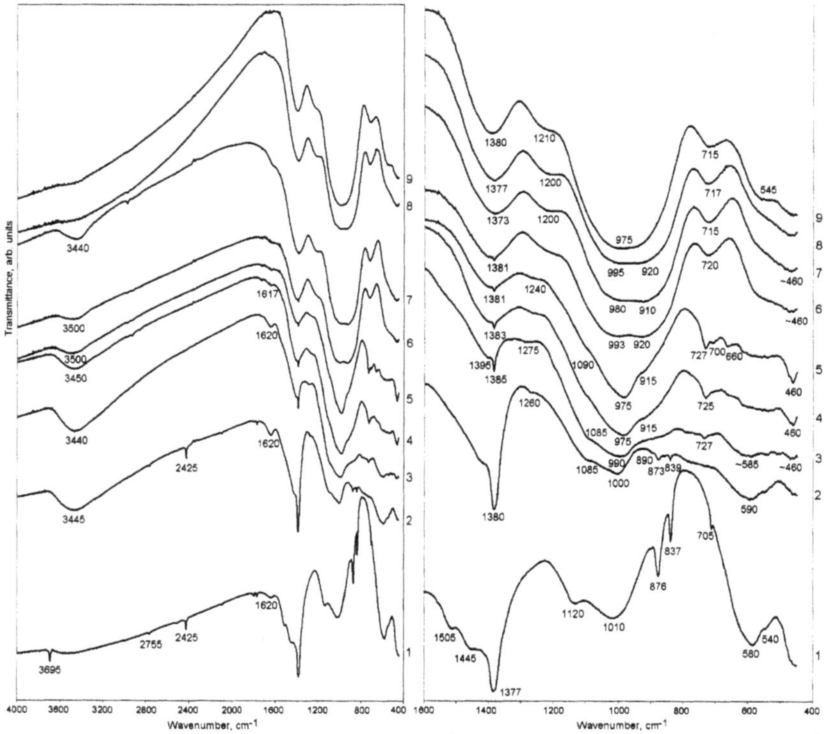

Figure 4. IR Spectra within the Range of 4000 to 400 cm^{-1} and Their Fragments of 1600 to 400 cm^{-1} of the Mixtures from SB4 Waste Surrogate (70 wt.%) and Frit 503-R4 (30 wt.%) Heat-Treated at Various Temperatures, °C: 1 – 115, 2 – 500, 3 – 700, 4 – 900, 5 – 1000, 6 – 1100, 7 – 1200, 8 – 1300, 9 – 1400.

CONCLUSION

As follows from both XRD and IRS data, reactions in the mixture of SB4 waste surrogate and Frit 503-R4 are completed within the temperature range between 700 and 900 °C when residual volatile anions are removed. This result is consistent with our previous study.[5] Nevertheless, minor intermediate pyroxene or sodalite-type phase is still present even after heat-treatment of the mixture at 1100 °C. At higher temperatures it is dissolved in borosilicate melt and the only crystalline phase in both slowly cooled (annealed) and quenched samples of the melt with 60 wt.% waste loading is spinel.

Phase formation reactions in the mixture at 70 wt.% waste loading are completed within the same range (700 to 900 °C) as in the mixture at 60 wt.% waste loading. However, very high melt viscosity requires much higher temperatures for homogenization and complete homogenization may be achieved at a temperature of ~1400 °C.

In the whole, phase formation reactions in the mixtures of SB4 waste surrogate and Frit 503-R4 go through formation of intermediate phases (pyroxene or sodalite-type, nepheline, low temperature spinel variety) and are completed within the range between 1000 °C and 1100 °C.

ACKNOWLEDGEMENTS

The work was performed under contract "CCIM Testing to Evaluate Processing Behavior of High Waste Loading Feeds" supported by US DOE Office of Environmental Management. Special thanks to Mr. K.D. Gerdes for funding and interest to work.

REFERENCES

[1]S.A. Dmitriev, S.V. Stefanovsky, *Management of Radioactive Waste*, Moscow (2000).

[2]S.L. Marra, R.J. O'Driscoll, T.L. Fellinger, J.W. Ray, P.M. Patel, J.E. Occipinti, DWPF Vitrification – Transition to the Second Batch of HLW Radioactive Sludge, *Waste Management '99 Conference*, Tucson, AZ, February 28 – March 4, 1999. ID 48-5.

[3] C. C. Herman, Summary of Results for Expanded Macrobatch 3 Variability Study, WSRC-TR-2001-00511 (2001).

[4]A.P. Kobelev, S.V. Stefanovsky, V.V. Lebedev, M.A. Polkanov, O.A. Knyazev, J.C. Marra. Cold Crucible Vitrification of the Savannah River Site SB2 HLW Surrogate at High Waste Loading, *Glass Technol. Part A*. **49**, 307-312 (2008).

[5] O.I. Stefanovsky, S.V. Stefanovsky, A.A. Akatov, J.C. Marra, Mechanisms of Phase Formation in the Vitrification of High-Ferrous Savannah River Site SB2 HLW Sludge Surrogate, *Waste Management 2009 Conference*, Phoenix, AZ, Match 1-5, 2009. ID 9300.

[6] J.C. Marra, "Sludge and Glass Compositions for Cold Crucible Induction Melter (CCIM) Testing – Sludge Batch 4," SRT-MST-2007-00070, Savannah River National Laboratory (2007).

[7] K. Nakamoto, "Infrared Specra of Inorganic and Coordination Compounds", John Wiley & Sons, New York, London (1962).

[8] L.V. Vilkov, Y.A. Pentin, "Physical Methods of Examination in Chemistry" (Russ.) Moscow (1987).

[9] I.I. Plusnina, "Infrared Spectra of Minerals," MGU, Moscow (1977).

[10] V.A. Kolesova, Vibrational Spectra and the Structure of Alkali Borate Glasses, *Glass Physics and Chemistry* (Russ.) **12** [1] 4-13 (1986).

[11] "Infrared Spectra of Alkali Silicates" (Russ.), A.G. Vlasov and V.A. Florinskaya (eds.), Leningrad (1970).

[12] S.V. Stefanovsky, I.A. Ivanov, A.N. Gulin, IR and EPR Spectra of Aluminoborosilicate and Aluminophosphate Glasses Simulating Vitrified Radioactive Waste, *Glass Physics and Chemistry* (Russ.) **17** [1] 120-125 (1991).

DEVELOPMENT OF WASTE FORMS FOR RADIOACTIVE IODINE

Terry J. Garino, Tina M. Nenoff, James L. Krumhansl and David X. Rademacher
Sandia National Laboratories
Albuquerque, New Mexico, USA

ABSTRACT

Radioactive iodine, ^{129}I is present in spent nuclear fuel and is of particular concern due to its extremely long half-life and its effects on human health. In the spent fuel reprocessing scheme under consideration, the iodine is released in gaseous form and collected using Ag-loaded zeolites, to form AgI either directly in the zeolite or with subsequent reaction with silver. We have investigated the use of low temperature-sintering glass powders mixed with either AgI or AgI-zeolite to produce a stable waste form that can be processed at 500°C, where AgI volatility is low. These mixtures can contain up to 20 wt% crushed AgI-zeolite and up to 50 wt% AgI. Crystallization of the BiZnB-oxide glass results in the encapsulation of the AgI, retaining it as a separate phase. The AgI-glass mixture was found to have the highest Iodine leach resistance in these initial studies.

INTRODUCTION

Radioactive iodine (^{129}I, half-life of 1.6×10^7 years) is generated in the nuclear fuel cycle. Due to its long half-life and its effects on human health, safe, long-term storage of ^{129}I is of particular concern. As part of the DOE/NE Advanced Fuel Cycle Initiative (AFCI), the separation of ^{129}I from spent fuel during fuel reprocessing is being studied. Specifically, in the process under consideration, gas containing ^{129}I vapor is passed over a bed of silver-exchanged zeolite, such as Ag-mordenite (Ag-MOR), to capture the ^{129}I. The captured ^{129}I can then be either (1) converted to AgI in the zeolite or (2) later driven off the zeolite and directly reacted with silver to produce AgI. Silver iodide (melting point = 552°C and density = 5.675g/cm^3) has a very low solubility in water, 1.3×10^{-8} M at 20°C, which makes it a good candidate for long term storage. However, the AgI, whether in the zeolite or not, must be incorporated into a dense solid waste form prior to storage.

One issue with using AgI to contain ^{129}I is its relatively high vapor pressure at moderate temperatures[1] (see Figure 1), which limits the thermal processing that can be done on a AgI-containing waste form without excessive loss of ^{129}I (>0.1%). Because of this, immobilization using borosilicate glass, for example, is not feasible due to excessive iodine loss at the required processing temperature. Other, low-melting, glasses such as vanadium and lead oxide glasses[2,3], do not meet chemical stability requirements. Another possibility that has recently been investigated is forming a glass using AgI and $Ag_4P_2O_7$.[4-6] In this case, the glass was melted at 500°C and was shown to have low solubility due to the formation a protective AgI surface layer after exposure to water. However, this approach may not be applicable to AgI-containing zeolites. The use of grout to contain AgI-MOR has also been investigated, but possibilities of carbonate release with time limit its effectiveness as a long term storage material.[7]

In this work, a new approach was investigated for forming ^{129}I-containing waste forms. This approach involved mixing either AgI or AgI-zeolite powder with a glass powder that can be sintered to high density at 500°C. Since the glass is not melted, a more refractory and therefore more chemically stable glass can be used. A commercially available bismuth oxide-based glass was chosen due to the low solubility of bismuth oxide in aqueous solution at pH>7.[8] In this approach, waste forms would be produced by crushing the AgI or the AgI-zeolite, mixing it with the glass powder, pressing the mixture into billets and then heating the billets to 500°C to densify the waste form. In this work, the feasibility of such a process was demonstrated and the materials produced were analyzed using powder X-ray diffraction, simultaneous Thermal Gravimetric and Differential Thermal Analyses (TGA/DTA) and

Scanning Electron Microscopy (SEM). An aqueous leaching study was also performed at 90°C on crushed samples of the AgI-zeolite and AgI mixed with the Bi-based glass.

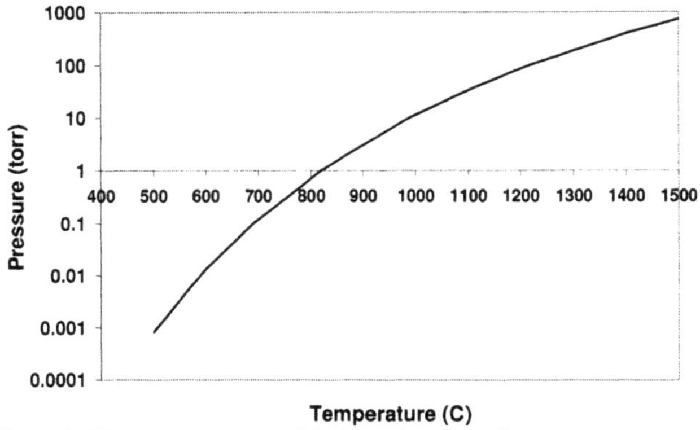

Figure 1. The vapor pressure of AgI as a function of temperature.

EXPERIMENTAL

Silver iodide (Apache Chem. Co.) was mixed with aluminum oxide powder (A3000FL, Alcoa Co.) and then analyzed using thermal gravimetric analysis at a heating rate of 10°C/min to determine when volatility of AgI begins to be significant. Similar samples containing 20 wt% AgI and a total mass of 5g were pressed into 2.5 cm diameter disks. Disks were heated to 500°, 550° or 600°C for 1 hr and their mass was measured before and after to determine the extent, if any, of AgI loss at that temperature.

The zeolite used was Ag-mordenite (Ag-MOR, Type AG900, Ionex Corp.). Since iodine vapor-loading of Ag-MOR requires a fairly involved pretreatment of the Ag-MOR to ensure that no elemental iodine is adsorbed,[9,10] a simpler loading technique was used in this initial study. The Ag-MOR was soaked for 1 hr at room temperature in a 10 wt% aqueous KI solution. After rinsing with deionized water and drying, the Ag-MOR turned yellow, indicating the presence of AgI. A ~20% mass increase was measured after the KI loading procedure. Powder X-ray diffraction analysis confirmed the presence of AgI on the loaded Ag-MOR.

The low sintering temperature glass used in this work, referred to hereafter as Glass A, was a bismuth-zinc-boron oxide glass that is available commercially (Ferro Corp., Cleveland, OH). This glass sinters to high density after only 15 min at 500°C and crystallizes during the sintering process. Manufacturer's properties of Glass A are given in Table 1. Several other glass compositions were also investigated but gave inferior results.

Table 1. Properties of Glass A (provided by Ferro Corp.).

Property	Value
Composition	BiZnB oxide
Recommended Sintering Conditions	500°C for 15 min
Coefficient of Thermal Expansion	8.8×10^{-6}/°C
Density	5.71 g/cm^3
Mean Particle Size	8 μm

Two types of samples were made using a mixture of Glass A and AgI. In the first type, the glass powder was simply mixed by mortar and pestle with AgI powder in equal parts by mass. The mixture was then dry pressed in a cylindrical steel die (2.5 cm diameter) at 70 MPa. The second type of sample contained the AgI-MOR that had been KI solution loaded. A mixture of 20 wt% AgI-MOR ground to -400 mesh was mixed by mortar and pestle with Glass A and then pressed into a billet. An initial sample of this mixture was heated in air to 500°C in a furnace equipped with a high temperature imaging system that allowed the shrinkage of the sample to be monitored *in situ*. The samples were then heated in air for up to 3 hr at 500°C. After sintering, the samples were analyzed using powder X-ray diffraction, TGA/DTA, and SEM using a back-scattered electron detector and energy dispersive spectroscopy (EDS). Also, some of each sample was used for solubility testing. A sample was ground to a 100 to 200 mesh fraction and placed in a PTFE container (Parr Instrument Co.) with deionized water. The solution was heated at 90°C for 7 days, in accordance with the Product Consistency Test (PCT).[11] The amount of dissolved I^- was determined using an I^--specific electrode (Orion 4Star, Thermo Scientific). For comparison, a sample of AgI-Ag phosphate glass (53.8 wt% AgI heated to 500°C for 3 hr) was prepared in accordance with reference 6 and was also leach tested.

RESULTS AND DISCUSSION

The TGA mass loss of a mixture of 10 wt% AgI with aluminum oxide during heating at 10°C/min in air is shown in Figure 2. Mass loss due to AgI volatilization begins at around 800°C, at ~1 torr vapor pressure (see Figure 1).

During waste form processing, the amount of AgI that vaporizes will depend not only on the maximum temperature but also on the time at temperature, as well as factors such as sample geometry (surface area to volume ratio) and AgI particle size. For example, a pressed disk 2.5 cm in diameter containing 20 wt% AgI mixed with alumina was heated for 1 hr and either 500°, 550° or 600°C. The sample heated to 500°C lost <0.1% of the initial AgI mass whereas those heated to the higher temperatures lost 0.3 and 0.5%, respectively. Therefore, it was concluded that heating to 500°C should result in acceptable AgI loss for times on the order of several hours for large samples size and that heating to 550°C for tens of minutes should also be acceptable.

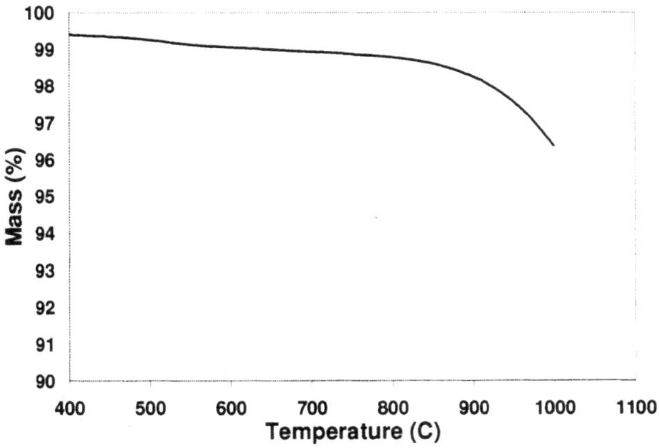

Figure 2. The TGA mass loss for a mixture of AgI (10 wt%) with alumina heated in air at 10°C/min.

The shrinkage behavior during heating at 5°C/min of a sample of 80 wt% Glass A and 20 wt% AgI-MOR, ground to -400 mesh, is shown in Figure 3. Shrinkage due to viscous sintering of the glass matrix begins ~400°C and reaches 20% linear shrinkage by 500°C. Little additional shrinkage was observed during the hold at 500°C. The Archimedes density of a sample held for 1 hr at 500°C was 4.4 g/cm^3 with <1% open porosity. When mixtures containing more than 20% AgI-MOR were heated to 500°C, the samples contained high levels of open porosity which is undesirable for chemical durability. When AgI was mixed directly with the Glass A powder, up to 50 wt% AgI could be added while still reaching essentially full density at 500°C (measured density of 5.43 g/cm^3 with <1% open porosity). This difference could be due to the finer particle size of the AgI as compared to the ground AgI-MOR. Mixtures of Glass A with more coarsely ground AgI-MOR could not be sintered to high density at even 20 wt%.

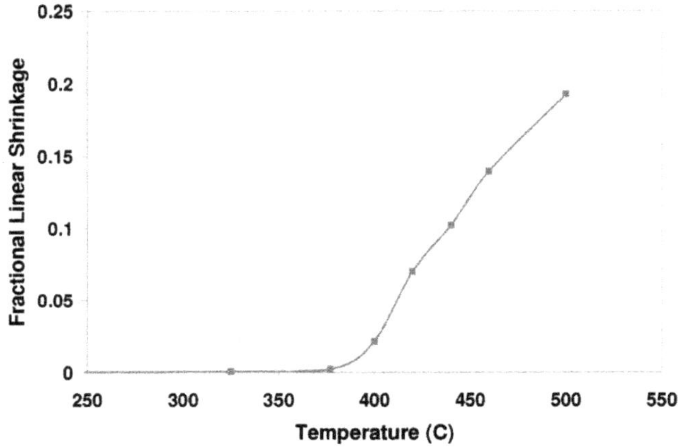

Figure 3. The shrinkage during heating at 5°C/min of a mixture of 20 wt% AgI-MOR (-38 μm) with Glass A.

Figure 4 shows the powder X-ray diffraction pattern of the 50 wt% AgI - 50 wt% Glass A sample that was heated to 500°C for 3 hr. The two primary phases identified were AgI and a bismuth oxide phase, either γ-Bi_2O_3, or more likely $Bi_{38}ZnO_{58}$.[12] The microstructure of the sample is shown in the SEM image in Figure 5. The AgI particles are ~25μm in size and are surrounded by a matrix of the crystallized Glass A, the lighter regions of which are the bismuth oxide phase. The matrix phase has some small cracks that could be the result of thermal expansion mismatch between the matrix and the AgI; the low temperature phase, β-AgI, has a negative coefficient of thermal expansion.[13] These cracks are most likely the cause of the small amount of open porosity measured using the Archimedes method. The same phases, with much less AgI, were found by XRD in the AgI-MOR-Glass A sample as shown in Figure 6. No mordenite peaks were observed indicating that it becomes amorphous during the heat treatment.

The microstructure of this material is shown in Figure 7. The dark particles are the remnants of the mordenite, since the EDS analysis indicated the presence of silicon and aluminum in these areas. They are roughly 10 μm in size. The rest of the microstructure consists of the crystallized Glass A matrix, with the bright white small particles being the bismuth oxide phase, again as indicated by EDS. The EDS spectra also had peaks for Ag and I. A map of the location of the Ag and I indicated that they were spread throughout the microstructure, not just in the mordenite regions.

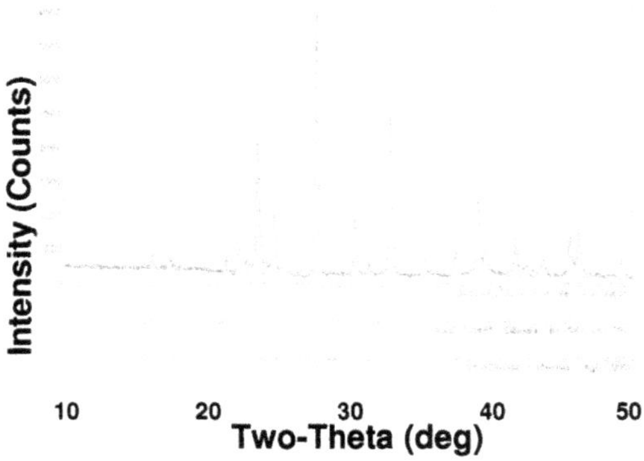

Figure 4. Powder XRD pattern of AgI (50 wt%) - Glass A mixture heated to 500°C for 3 hr.

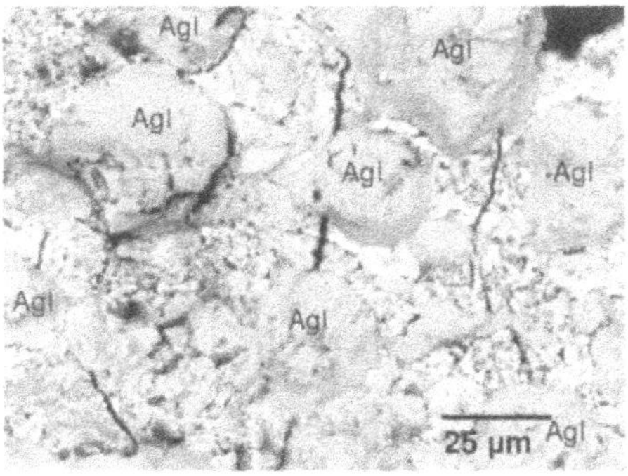

Figure 5. SEM image of the 50 wt% AgI/50 wt% Glass A sample after 3 hr at 500°C.

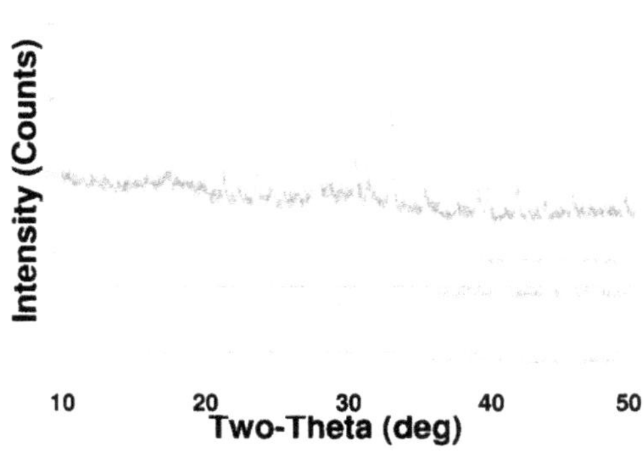

Figure 6. Powder XRD pattern of a AgI-MOR (20 wt%) mixture with Glass A heated to 500°C for 1 hr.

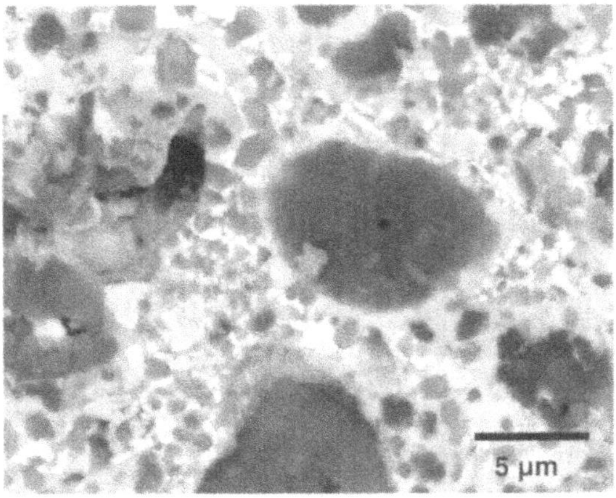

Figure 7. SEM microstructural image of the AgI-MOR/ Glass A sample.

The TGA/DTA results on the sintered AgI-Glass A sample are shown in Figure 8. The TGA (green curve) indicated that the material did not undergo significant mass loss due to AgI volatilization until above 600°C, with very rapid loss above 900°C. During heating to 1000°C, peaks in the DTA (blue curve) occurred at ~150°C where AgI undergoes a change in crystal structure and also near 550°C, corresponding to the crystallization of AgI. These thermal events could not be resolved in the 20 wt% AgI-MOR sample, due to the lower concentration of AgI, see Figure 9. This zeolite-containing sample began to lose mass (due to AgI vaporization) above 600°C with rapid loss occurring >800°C.

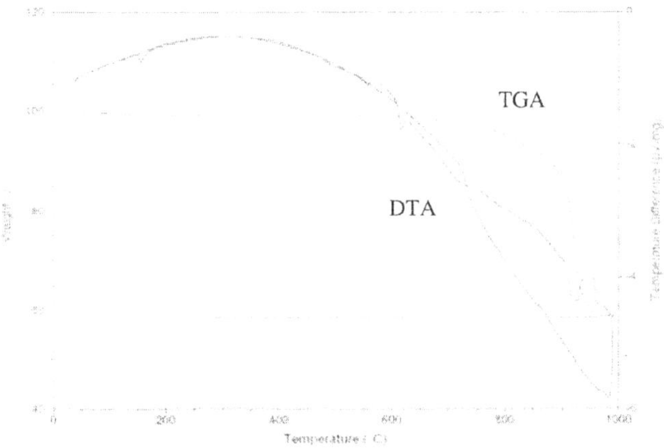

Figure 8. TGA(green)/DTA(blue) curves for of AgI (50 wt%) - Glass A mixture heated to 500°C for 3 hr.

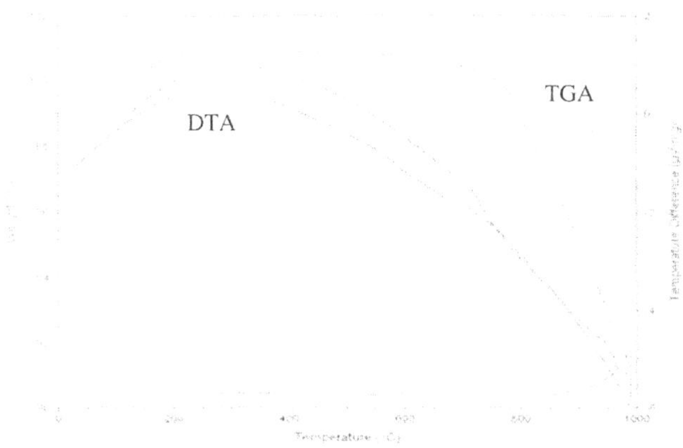

Figure 9. TGA(green)/DTA(blue) curves for AgI-MOR (20 wt%) mixture with Glass A heated to 500°C for 1 hr.

The results of the leaching study indicated that the AgI-Glass A sample produced a leachant with an I^- concentration of 0.6 x 10^{-6} mol/liter. The I^- concentration in the leachant for the 20 wt% AgI-MOR/Glass A sample was much higher, 340 x 10^{-6} mol/liter. This was most likely an artifact of the method used to iodine load the Ag-MOR. Some soluble iodide was most likely sorbed to the bulk surface of the zeolite from the KI solution treatment. For comparison, the I^- concentration for the AgI-Ag phosphate glass was 72 x 10^{-6} mol/liter, significantly higher than that for the AgI-Glass A sample.

CONCLUSIONS

The feasibility of using a low sintering temperature glass to encapsulate silver iodide, either alone or in a zeolite, has been demonstrated. A commercial BiZnB oxide glass powder is mixed with up to 50 wt% AgI or 20 wt% of AgI-MOR, pressed and densified at 500°C, resulting in minimal AgI loss due to volatilization. Initial leach testing for iodine solubility indicated that levels below 1 micromole/liter can be achieved. Further research is in progress to determine solubility of the other components. In addition, other low temperature sintering glasses are being considered.

ACKNOWLEDGEMENTS

The authors acknowledge the US DOE/NE-AFCI program for funding support. The authors thank Gary Zender for the SEM analysis. This work was performed at Sandia National Laboratories, Albuquerque, NM. Sandia is a multiprogram laboratory operated by Sandia Corporation, a Lockheed Martin Company, for the United States Department of Energy's National Nuclear Safety Administration under Contract DE-AC04-94AL85000.

REFERENCES

[1]CRC Handbook of Chemistry and Physics, 61st Ed., R.C. Weast ed., 1980.

[2]T. Nishi, K. Noshita, T. Naitoh, T. Namekawa, K. Takahashi, M. Matsuda, Applicability of V_2O_5-P_2O_5 Glass System for Low-Temperature Vitrification, *Mat. Res. Soc. Symp. Proc.* **465**, 221-8 (1999).

[3]D. Perera, E. Vance, R. Trautman, B. Begg, Current Research on I-129 Immobilization, *Proc. of WM'04 Conference*, WM-4089 (2004).

[4]H. Fujihara, T. Murase, T. Nishi, K. Noshita, T. Yoshida, and M. Matsuda, Low Temperature Vitrification of Radioiodine Using AgI-Ag_2O-P_2O_5 Glass System, *Mater. Res. Soc. Symp. Proc.* **Vol. 556**, 375-382. (1999).

[5]K. Noshita, T. Nishi, T. Yoshida, H. Fujihara and T. Marase, "Vitrification Technique of Radioactive Waste Using AgI-Ag_2O-P_2O_5 Glass System", in *InternationalConference Proc. Radioactive Waste Management and Environmental Remediation*, pub., ASME, New York, 107-112 (1999).

[6]T. Sakuragi, T. Nishimura, Y. Nasu, H. Asano, K. Hoshino and K. Iino, Immobilization of Radioactive Iodine Using AgI Vitrification Technique for the TRU Wastes Disposal: Evaluation of Leaching and Surface Properties, *Mater. Res. Soc. Symp. Proc.* **Vol. 1107** (2008).

[7]R. D. Scheele, C. F. Wend, W. C. Buchmiller, A. E. Kozelisky, R. L. Sell, "Preliminary Evaluation of Spent Silver Mordenite Disposal Forms Resulting from Gaseous Radioiodine Control at Hanford's Waste Treatment Plant," Battelle - Pacific Northwest Division, Richland, Washington, December 2002.

[8]C.F. Baes and R.E. Mesmer, **The Hydrolysis of Cations**, John Wiley and Sons, Inc., 1976.

[9]E. Vance and D. Agrawal, X-Ray Studies of Iodine Sorption in Some Silver Zeolites, *J. Mater. Sci.* **17** 1889-1894 (1982).

[10]R.D. Scheele, L.L. Burger, and C.L. Matsuzaki., "Methyl Iodide Sorption by Reduced Silver Mordenite," PNL-4489, Pacific Northwest Laboratory, Richland, Washington (1983).

[11]Standard Test Methods for Determining Chemical Durability of Nuclear, Hazardous, and Mixed Waste Glasses and Multiphase Glass Ceramics: The Product Consistency Test (PCT), Designation: C 1285 – 02, ASTM Int., West Conshohocken, PA, (2008).

[12]J. P. Guha, S. Kunej, and D. Suvorov, "Phase Equilibrium Relations in the Binary System Bi_2O_3-ZnO," J. MATER. SCI. **39** 911– 918 (2004).

[13]G. Harvey and N. H. Fletcher, "Thermal Expansion of β-Silver Iodide at Low Temperatures," J. Phys. C: Solid St. Phys. **13** 2969-73 (1980).

$CaMOO_4$ IN A MOLYBDENUM RICH BOROSILICATE GLASS-CERAMIC: A SPECTROSCOPIC STUDY

Clément Mendoza ; Gérard Panczer ; Dominique de Ligny
Laboratoire de Physico-chimie des matériaux luminescents Université de Lyon
Villeurbanne, France

Isabelle Bardez-Giboire ; Sophie Schuller
DEN/VRH/DTCD/SECM/LDMC CEA Valrhô Marcoule
Bagnols-sur-Cèze, France

Sylvain Peuget
DEN/VRH/DTCD/SECM/LMPA CEA Valrhô Marcoule
Bagnols-sur-Cèze, France

ABSTRACT

Calcium molybdate, $CaMoO_4$ (powellite) is the main micro phase crystallizing during the cooling of molybdenum rich borosilicate glasses studied for the development of future containment matrices. The composition and the structure of the vitreous and the crystalline phases of such a glass-ceramic is determinant for the long-term behaviour understanding of such a matrix. In this study Eu and Nd rare earth elements (REEs) were used as fission product (FP) and minor actinide (MA) surrogates. Raman spectroscopy based on oriented natural powellite sample spectra indicates that the crystallization direction of the tetragonal powellite dendritic crystals in the glass-ceramic was along the c axis. Microprobe and photoluminescence analyses reveal that the REEs are preferentially incorporated in the powellite micro phases associated with Sr and Na substitutions. A $Ca0_{.76}Sr_{0.1}Na_{0.07}Eu_{0.01}La_{0.02}Nd_{0.02}Pr_{0.02}MoO_4$ ceramic was therefore synthesized as a polycrystalline simple model and allowed to demonstrate the formation of a $(CaMoO_4)_{0.76}$-$(SrMoO_4)_{0.1}$-$(NaREEMoO_8)_{0.07}$ solid solution.

INTRODUCTION

In order to increase the Fission Products (FP) and Minor Actinides (MA) rates (> 18.5 w%) resulting from the UOX type used fuel withdrawal in the containment glasses, some new glass compositions (HTC) are developed[1].

One of the loading rate limitations came from the molybdenum oxide low solubility (on the order of 1%) in the sodium borosilicate silica rich matrix. Beyond the solubility limit, a macroscopic liquid-liquid phase separation phenomenon conducts, during the cooling, to the formation of a low durability crystalline alkaline and alkaline-earth molybdate complex comonly named "yellow phase"[2,3-3]. Because of the waste loading rate limitation (< 18.5 w%) and the optimization of the precursors chemical reactivity conditions, those phases are avoided in the present glasses. At lower temperature in the bulk, other crystallization processes could be produced independently from the phase separation phenomenon[4].

The study of the nucleation and growth kinetics allows to predict the long-term thermic stability of the R7/T7 type glasses[5,6].

A liquid-liquid microscopic phase separation mechanism appears in glasses with high molybdenum oxide and low silica contents, studied for the containment of the wastes resulting from the UMo fuels withdrawal[7] and leads to the formation of a glass-ceramic[8]. This nucleation-growth mechanism[9] creates at high temperature some separated phases that crystallise during the cooling to form alkaline and alkaline-earth phosphates, zircons and neodymium rich calcium molybdates[10] crystalline phases.

In addition to the technological feasibility, the produced glasses quality, their chemical durability and the glassy matrix thermal stability, the behaviour under self-irradiation is one of the specifications to take into account[11].

Those previous studies demonstrate the possibility of powellite crystallization in a SON68 type[5,6] or UMo[10] type glasses and the conditions of such crystallization but not the real composition of the crystalline phase. The possibility of a rare earth element (REE) incorporation, the Nd, in the powellite structure is shown by Henry[10] in an UMo type glass but there's no data for the SON68 type glass that contains several REEs, FP and MA capable of entering in the powellite structure. Those elements are not the only ones susceptible to form molybdates. Elements such as Sr and Na are good candidates to substitute a Ca and all those substitutions have a direct impact on the crystalline structure during the crystal formation (increasing of the cell parameters) and during its lifetime (disintegration of the MA and FP incorporated). Also crystallinity due to cation ordering can evolve with thermal history[12]. A better understanding of the crystalline phase is thus fundamental to know its durability and predict its behaviour in the glass-ceramic.

Furthermore, with increasing FP and MA quantities in glasses, rare earth elements and actinides substitutions in the calcium molybdate phase could increase dramatically.

This study, prior to the study of the crystalline phases behaviour under irradiation, will focus on the spectroscopic characterisation of the REEs substituted calcium molybdate type crystalline phases by photoluminescence and Raman spectroscopy. For that purpose, a model system derivated from the SON68 glass was defined and the model glass-ceramic was obtained after an annealing at 800 °C. The composition of the crystals was determined with a microprobe. A ceramic phase with the main elements found in this analysis was synthesised as a model system for the glass-ceramic crystal phase for the future study on the crystalline phase behaviour under irradiation.

SAMPLES

Glass

For this study, the SM glass, a non radioactive composition derivated from the SON68 glass was synthesised (table I). Some elements such as Cd, Cr and Ni were removed in order to produce a less opaque glass for a better analysis with optical spectroscopic techniques such as Raman spectroscopy and photoluminescence. The Eu was added (in the place of Ce) and the concentration of Nd was increase because of their luminescent properties and their role as surrogate of the FP and the MA. Finally, the Mo concentration has been increased to enhance the $CaMoO_4$ crystallization.

Table I. SM Glass composition.

Oxides wt%	SM glass	Precursors
SiO_2	46,42	$Si0_2$
B_2O_3	14,32	H_3BO_3
Na_2O	10,35	Na_2CO_3
Al_2O_3	5,07	Al_2O_3
CaO	4,12	$CaCO_3$
Li_2O	2,02	Li_2CO_3
Fe_2O_3	3,07	Fe_2O_3
ZnO	2,56	ZnO
P_2O_5	0,29	$AlPO_4$
SrO	0,35	$Sr(NO_3)_2$
ZrO_2	2,79	ZrO_2
MoO_3	2,54	MoO_3
MnO_2	0,39	$Mn(NO_3)_2$, $4H_2O$
Cs_2O	1,11	$CsNO_3$
BaO	0,63	$Ba(NO_3)_2$
Y_2O_3	0,20	Y_2O_3
La_2O_3	0,95	La_2O_3
Nd_2O_3	1,67	Nd_2O_3
Pr_2O_3	0,48	Pr_6O_{11}
Ag_2O	0,03	Ag_2O
SnO_2	0,02	SnO
TeO_2	0,23	TeO_2
Eu2O3	0,39	Eu_2O_3
Total	100,00	

The glass (~600 g batch) was prepared using the reagent grade powder listed in the table I melted under air in a platinum/rhodium/yttrium crucible at 1200 °C for 3 h. It was cooled on a metallic plate for a speed cooling about 10^3 °C/min.

Glass-ceramic

Orlhac[5] showed that two morphology of CaMoO4 can crystallize in a SON68 type glass. A needle shaped form appears between 690 and 810 °C and a small stocky bipyramidal form appears between 810 and 880 °C The crystals must be large enough (>3x3 μm) to be easily analysed with a microprobe (in order to determine their composition) and with Raman and microluminescence spectroscopies. Based on TTT diagrams[5], the glass-ceramic SR01 was then obtained from the S glass after a 64 h annealing at 800 °C. This annealing time is necessary to obtain large needle shaped crystals (100-1000 μm long and 10-20 μm large), the size of the bipyramids could not be increased enough (Figure 1). This glass-ceramic in thus composed by SR01 needle shaped crystals surrounding by the SR01 glassy matrix.

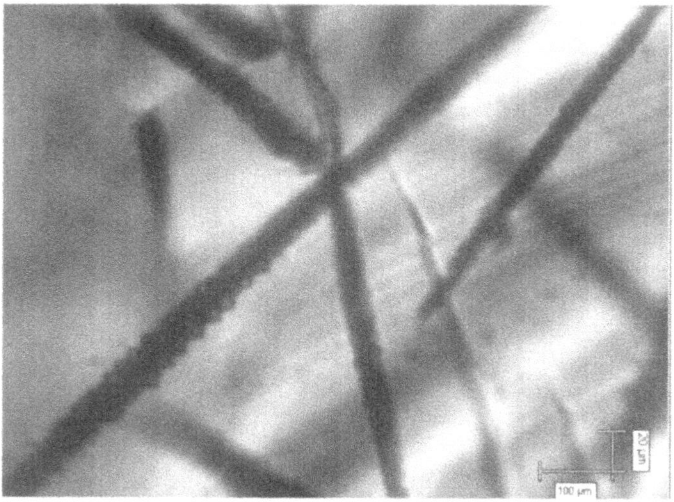

Figure 1. Optical image of the SR01 glass-ceramic (x50 magnification) with needle shaped crystals.

Ceramic

A $Ca_{0.76}Sr_{0.1}Na_{0.07}Eu_{0.01}La_{0.02}Nd_{0.02}Pr_{0.02}MoO_4$ ceramic (CM07) was synthesised from the SR01 crystals composition (Table II) as a simple polycrystalline model.

Table II. Molar composition of the SR01 crystals and the CM07 ceramic, the number in brackets is the error on the last decimal.

Molar concentration	SR01 crystals WDS analysis	CM07 elaboration composition	CM07 EDX analysis
O	4.00	4.00	4.00
Ca	0.69 (1)	0.76	0.83
Mo	0.72 (2)	1.00	0.97
Sr	0.10 (2)	0.10	0.10
Na	0.088 (4)	0.07	0.07
Nd	0.03 (1)	0.02	0.02
Pr	0.024 (1)	0.02	0.04
La	0.021 (8)	0.02	0.02
Eu	0.007 (1)	0.01	0.01
B	0.45 (2)	0.00	0.00
Others	0.06	0.00	0.00

MoO_3, $CaCO_3$, Na_2CO_3, $SrCO_3$, Pr_6O_{11}, Nd_2O_3, Eu_2O_3 and La_2O_3 oxide powders (Alfa Aesar 99.0%, Fluka 99.9%, Prolabo 99.0 and 99.995%, CERAC 99.9% and Sodipro 99.9%) were used as reactants. Powders were intimately mixed at solid state and calcined at 800 °C during 6 h under air. The resulting powder was then grounded in a planetary crusher under water during 1 hour. Finally, pellets were obtained by a cold uniaxial pressing at 130 MPa and ceramics were sintered at 1150 °C during 3h under air.

The X-Ray diffraction were used after each step in order to control the nature of the phase obtained and electronic microscopy analysis (SEM/EDX) was done on the final sintered pellet in order to determine the microstructure and the real composition (Table II) of the phases present in the ceramic. The EDX analysis was performed using wollastonite, jade, metal molybdenum, SrF_2, NdF_3 and LaB_6 standards. Eu and Pr concentrations were semi quantitatively estimated. The ceramic is homogenous, dense (densification ratio > 98%) with a little porosity (10% surfacic porosity).

Reference materials

The Na_2MoO_4 reference is a commercial ALDRICH CAS 7631-95-0 powder. The CaMoO4 reference ceramic was synthesized with the same process as the CM07 one. XRD, SEM/EDX and Raman spectroscopy techniques confirm that this ceramic is pure and well-crystallized CaMoO4. The CM06 reference ceramic composition is: $Ca_{0.86}Na_{0.07}Eu_{0.01}La_{0.02}Nd_{0.02}Pr_{0.02}MoO_4$. It's the same as CM07 but without Sr. All the synthesis process was identical to the CM07 one. XRD, SEM/EDX and Raman spectroscopy techniques confirm that this ceramic is pure and is a well-crystallized $CaMoO_4$ type material.

The monocrystal used as reference for the orientation Raman spectra is a 9x8x5 mm natural automorph Indian powellite monocrystal.

EXPERIMENTAL SET-UP

Oriented Raman and luminescence spectra have been recorded at room temperature on a Renishaw RM1000 microspectrometer with a 1800 grooves per mm grating. The laser (532 nm –Laser Quantum-) was focused on the sample thought a x50 Olympus objective. The signal was collected through the same objective and a 10 µm open slit in a CCD cooled by Pelletier effect. Neither the laser nor the detector are polarized. The laser is horizontally oriented and the changes of polarization were simulated with a 90° rotation of the sample. The spectra where thus described using the Porto notation: x(yR)x, where the letters outside the brackets represent respectively the laser excitation and the signal collection directions. In our system, the emitted and the collected laser have the same direction. The letters inside the brackets represent respectively the laser and the detector polarization. (x,y,z) are used when the (a,b,c) crystalline axis are known, otherwise (X,Y,Z) are used. The non polarization of the detector is represented by the R letter.

The non-oriented Raman spectra have been recorded at room temperature on a Horiba Jobin Yvon Aramis microspectrometer with a 1800 grooves per mm grating. The laser (He/Ne 632.5 nm) was focused on the sample thought a x50 Olympus objective. This excitation wavelength allows us to avoid luminescence in the Raman spectra range.

RESULTS AND DISCUSSION

Phase characterisation

The CM07 SEM/EDX analysis (table II) shows that its composition is very close to the nominal one. The ceramic is homogeneous and the XRD spectrum corresponds to a powellite type material similar to the main crystal phase of the SR01 glass-ceramic (Figure 2).

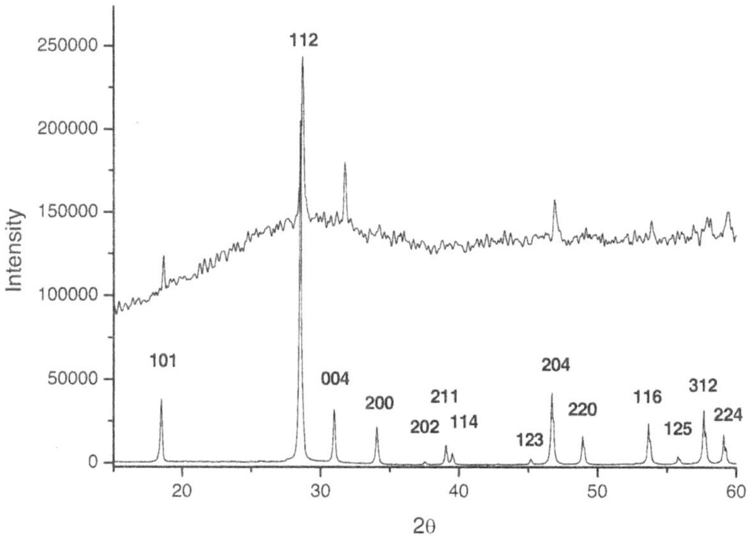

Figure 2. XRD spectra of the glass ceramic SR01 (top) and ceramic CM07 (bottom) samples. The (hkl) crystalline plans Bragg positions of the powellite are indicated.

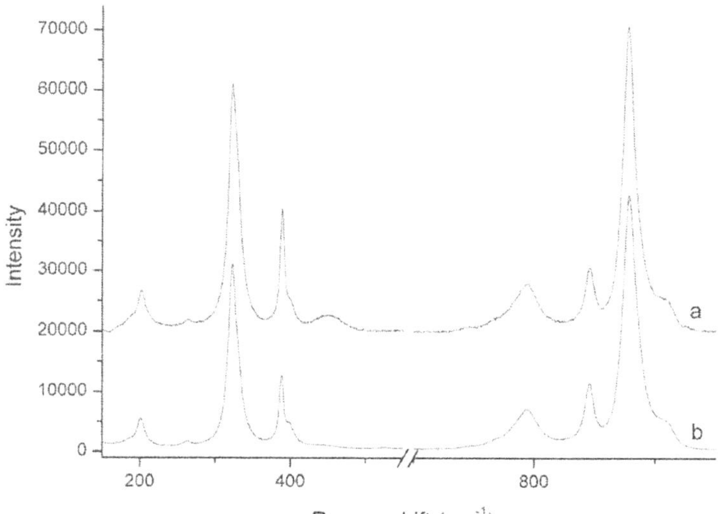

Figure 3. Raman spectra of (a) SR01 crystals and (b) CM07 ceramic.

The SR01 crystals and CM07 ceramic Raman spectra are almost the same (Figure 3). The slight differences (the broad band at 440 cm^{-1} in the SR01 crystal) could be explained by the presence of impurities in the SR01 crystals.

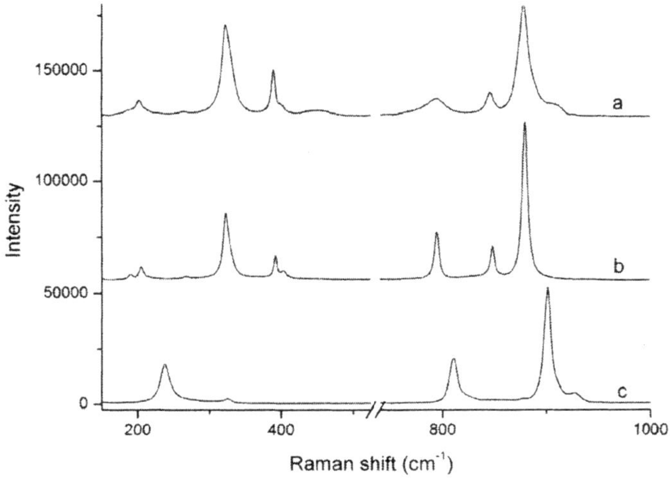

Figure 4. Raman spectra of (a) SR01 crystals and (b) $CaMoO_4$ ceramic and (c) Na_2MoO_4 powder.

As shown in Figure 4, the SR01 Raman spectrum fit well the one of the $CaMoO_4$ ceramic. The main differences are a shoulder at higher wave number for the main 880 cm^{-1} band, a slight shift at lower wave numbers of the 400 cm^{-1} bands and a band broadening (especially in the 750-900 cm^{-1} area). This broadening suggests a lower crystallinity of the SR01 crystals and the CM07 ceramic. This lack of crystallinity can be associated with cation substitution and disorder.

The SR01 crystals and the CM07 ceramic are quite similar to powellite type crystals and the differences found when their Raman spectra are compared to the one of a pure $CaMoO_4$ crystal could find an explanation with substitutions in a classic powellite structure.

REEs substitution and Na^+ compensation

The incorporation of rare earth elements (REEs) such as Nd is well developed in the LED[13] and laser crystals[14-16] researches. The 3+ charge of the REEs can be compensated by a mono-charged elements[13] such as Na^+, Li^+ or K^+ for a better integration of the REE in the crystalline structure.

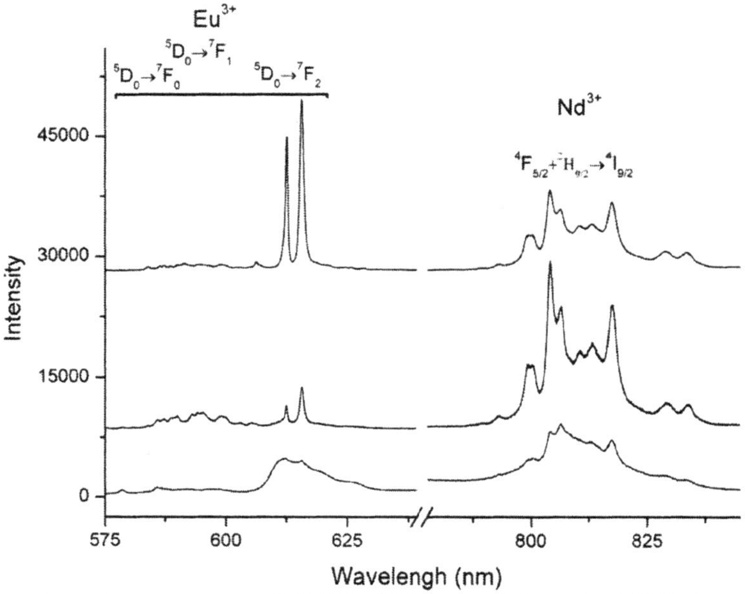

Figure 5. Luminescence spectra of CM07 (top) SR01 crystals (center) and SR01 glassy matrix (bottom) under 532 nm excitation.

In our case, several REEs could enter in the crystalline structure because of their presence in the SM glass before annealing. The luminescence of Eu^{3+} and Nd^{3+} can be seen in the SR01 glassy matrix and crystals (Figure 5) spectra but they are quite different. The crystals spectrum is more structured and this is characteristic of an incorporation of luminescent ion in a crystalline structure. The luminescence spectrum of the CM07 ceramic is very close to SR01 crystals one, the main difference being the luminescence of the $^7D_0 \rightarrow ^7F_2$ transition that is stronger in the ceramic.

In SR01 crystals (as well as in CM07), Na concentration is roughly the same as the sum of the REEs concentrations. The Na can crystallise as a sodium molybdate Na_2MoO_4 with a cubic or orthorhombic structure[17]. The Figure 4 shows a Raman spectrum of a Na_2MoO_4 ceramic. Even if the shoulder of the SR01 crystals 880 cm^{-1} band is in the same wave number range than the main Na2MoO4 band, the other bands from the sodium molybdate could not be seen on the SR01 crystals spectrum. This confirms the results of the XRD and SEM/EDX analysis of CM07, there's only one homogenous phase that crystallizes. The Na takes the place of a Ca in the powellite structure.

The hypothesis is that in those crystals, the Na^+ is the charge compensator of the REE^{3+} in the powellite structure, creating a $CaMoO_4$-$NaREEMo_2O_8$ solid solution. Such solid solutions were already the subject of Monte-Carlo simulations[18].

The effects of those substitutions on the crystalline structure could be seen in Figure 6 when comparing a CaMoO4 ceramic with the $Ca_{0.86}Na_{0.07}Eu_{0.01}La_{0.02}Nd_{0.02}Pr_{0.02}MoO_4$ (CM06) ceramic. The a cell parameter increases from 5.22 Å for a pure CaMoO4 ceramic to 5.24 Å for the CM06 ceramic and the c cell parameter from 11.43 to 11.47 Å. This unit cell dilatation isn't totally uniform (0.38 % along a axis and 0.35 % along c axis).

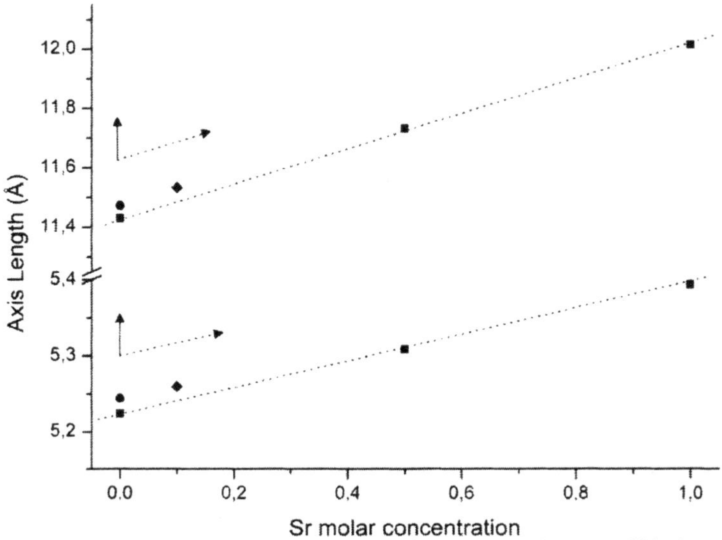

Figure 6. Variation of the a (bottom) and c (top) cell parameters. The squared black symbols represent different $Ca_xSr_{1-x}MoO_4$ crystals[19]. The round and the diamond-shaped symbols represent respectively the CM06 and CM07 ceramics. The shift caused by the incorporation of Na and REEs in the powellite structure is symbolized by the plain arrows, the shift caused by the incorporation of Sr by the doted arrows.

Sr substitution

The effect on the unit cell parameters of Sr incorporation in $Ca_xSr_{1-x}MoO_4$ solid solution can be see in Figure 6. The a and c cell parameters grow linearly with the Sr incorporation from respectively 5.22 and 11.43 for $CaMoO_4$ to 5.39 and 11.73 for $SrMoO_4$. This correspond to a dilatation of 3.3 % along the a axis and 2.6 % along the c axis.

In CM07, the dilatation caused by the Sr incorporation superposes the one caused by the REEs and Na incorporation. But when the contribution of the Na and the REEs is subtracted, the CM07 ceramic has a dilatation corresponding to a $Ca_{0.9}Sr_{0.1}MoO4$ composition. This confirms the expected composition and the EDX measurement.

Finally, the SR01 crystals as well as the CM07 ceramic, could be seen as a $(CaMoO_4)_{0.76}$ - $(SrMoO_4)_{0.1}$ - $(NaEuMoO_8)_{0.01}$ - $(NaLaMoO_8)_{0.02}$ - $(NaNdMoO_8)_{0.02}$ - $(NaPrMoO_8)_{0.02}$ solid solution.

Crystallization orientation

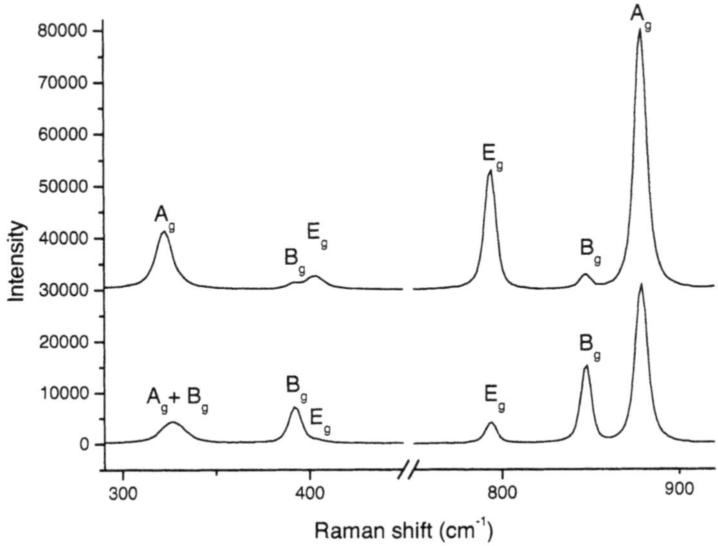

Figure 7. Oriented Raman spectra of a natural powellite monocrystal. Top to bottom: x(zR)x and z(xR)z.

The powellite crystallizes in the tetragonal I 4_1/a space group[20]. In order to identify their growth direction during devitrification and annealing, polarized Raman spectra were recorded. Each orientation was indexed according Porto notation[20] : crystallographic direction parallel to the laser beam and to the scattering light collection are indicated out of the brackets and direction of polarization, in the brackets (nb. In our case the scattered light is randomly collected and anoted R). The anisotropy can be seen on the Raman spectra of a oriented monocrystal (Figure 7). Depending on the laser polarization and the sample orientation, some vibrational modes will be extinct. Tallying with the conclusion of Porto and Scott[20], the B_g modes for a x(zR)x orientation, and the E_g modes for a z(xR)z orientation tend to be extinct. The B_g modes are enhanced for the z(xR)z orientation.

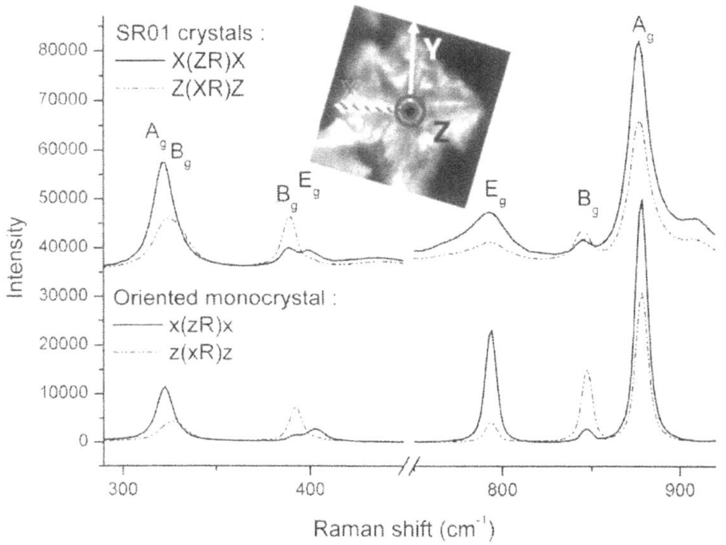

Figure 8. Comparison of the oriented Raman spectra of the SR01 crystals (top) and a $CaMoO_4$ natural monocrystal. For the SR01 crystals (optical image on top), the Z axis was defined along the growth axis.

In the SR01 glass-ceramic, the crystals are needle shaped. Because of the anisotropy of the crystalline structure, the crystallisation direction of the needles can have a direct impact on their behaviour in particular under irradiation.

In order to determine this direction, axis were arbitrary plotted on a transversal cut of a crystal for an oriented Raman spectroscopy study: the Z axis along the crystallization direction, X and Y perpendicularly to it (Figure 8). For the Z(XR)Z spectrum, the A_g modes decreased and the B_g modes enhanced as in a z(xR)z spectrum. This result is confirmed by the B_g modes extinction for the X(ZR)X spectrum which correspond to an x(zR)x spectrum. The arbitrary Z axis is the c axis in the tetragonal powellite cell. The X and Y axis equal to the a axis. Finally, the needles crystallize along the c axis. Dendritic crystals characterize crystallization constrained by diffusion processes.

CONCLUSION

The Eu^{3+} luminescence shows very little differences between the SR01 crystals and CM07 ceramic spectra but as this luminescence is very sensitive to the Eu environment, the CM07 ceramic can be considered as a good polycrystalline model for the glass-ceramic crystal phase. This ceramic is elaborated to be implanted with heavy ion beam to study the behaviour of the crystalline phase under irradiation.

Sr, Na and REEs substitute to Ca during the crystallization of $CaMoO_4$ in Mo rich borosilicate SON68 type glass. The final product can be defined as a $(CaMoO_4)_{0.76}$ - $(SrMoO_4)_{0.1}$ - $(NaEuMoO_8)_{0.01}$ - $(NaLaMoO_8)_{0.02}$ - $(NaNdMoO_8)_{0.02}$ - $(NaPrMoO_8)_{0.02}$ solid solution. While the goal is to increase the FP and MA quantities in the storage glass, further studies must be conducted to determine if, when increasing the REEs concentration in the glass, there is a limit in their incorporation in the $CaMoO_4$ type phase.

ACKNOWLEDGEMENT
This work received a financing from the Region Rhône-Alpes via the MACODEV cluster.

REFERENCES

[1]J. L. Dussossoy, S. Schuller, and T. Advocat, The formulation of new high level glasses with an increased waste loading charge, *International Congress of Glass XXI* (2007).
[2]R. J. Short, R. J. Hand, and N. C. Hyatt, Molybdenum in nuclear waste glasses. Incorporation and redox state *Mat. Res. Soc. Symp. Proc.*, **757**, 141-143 (2003).
[3]G. Calas, M. Le Grand, L. Galoisy, and D. Ghaleb, Structural role of molybdenum in nuclear glasses: an EXAFS study, *Journal of Nuclear Materials*, **322**, 15-20 (2003).
[4]X. Orlhac, C. Fillet, and J. Phalippou, Study of crystallisation mechanisms in the French nuclear waste glass, *Mat. Res. Soc. Symp. Proc.*, **556**, 263-270 (1999).
[5]X. Orlhac, Etude de la stabilité thermique du verre nucléaire. Modélisation de son évolution à long terme, Montpellier II, (1999).
[6]X. Orlhac, C. Fillet, P. Deniard, A. M. Dulac, and R. Brec, Determination of the crystallized fractions of a largely amorphous multiphase material by the Rietveld method, *Journal of Applied Crystallography*, **34**, 114-118 (2001).
[7]R. D. Quang, V. Petitjean, F. Hollebecque, O. Pinet, T. Flament, and A. Prod'homme, Vitritrifation of HLW produced by Uranium/Molybdenum Fuel Reprocessing in Cogema's Cold Crucible Melter *Waste management 2003 Symposium, Tucson, AZ* (2003).
[8]S. Schuller, O. Pinet, A. Grandjean, and T. Blisson, Phase separation and crystallization of borosilicate glass enriched in MoO_3, P_2O_5, ZrO_2, CaO, *Journal of Non Crystalline of Solid* **354**, 296-300 (2008).
[9]S. Schuller, A. Grandjean, O. Pinet, P. Perouty, and T. Blisson, Phase separation and crystallisation in UMo glass, *International Conference Atalante 2004, 21-24 juin 2004, Nîmes (France)* (2004).
[10]N. Henry, P. Deniard, S. Jobic, R. Brec, C. Fillet, F. Bart, A. Grandjean, and O. Pinet, Heat treatments versus microstructure in a molybdenum-rich borosilicate, *Journal of Non-Crystalline Solids*, **333**, 199-205 (2004).
[11]S. Peuget, J.-N. Cachia, C. Jégou, X. Deschanels, D. Roudil, V. Broudic, J. M. Delaye, and J.-M. Bart, Irradiation stability of R7T7-type borosilicated glass, *Journal of Nuclear Materials*, **354**, 1-13 (2006).
[12]D. De Ligny, G. Panczer, D. Caurant, and D. Neuville, Contribution of neodymium optical spectroscopy to the crystal growth study of a silicate apatite in a glassy matrix, *Optical Materials*, **30**, 1694-1698 (2008).
[13]J. Liu, H. Lian, and C. Shi, Improved optical photoluminescence by charge compensation in the phosphor system $CaMoO_4$:Eu^{3+}, *Optical Materials*, **29**, 1591-1594 (2007).
[14]E. Cavalli, E. Bovero, and A. Belletti, Optical spectroscopy of $CaMoO_4$: Dy^{3+} single crystals, *Journal of Physics-Condensed Matter*, **14**, (20), 5221-5228 (2002).
[15]L. H. C. Andrade, D. R. Ardila, L. B. Barbosa, J. P. Andreeta, M. S. Li, A. Brenier, Y. Guyot, and G. Boulon, Spectroscopic study of floating zone technique-grown Nd^{3+}-doped $CaMoO_4$, *Eur. Phys. J.-Appl. Phys*, **29**, 55-64 (2005).
[16]L. H. C. Andrade, M. S. Li, Y. Guyot, A. Brenier, and G. Boulon, Optical multi-sites of Nd^{3+}-doped $CaMoO_4$ induced by Nb^{5+} charge compensator, *Journal of Physics-Condensed Matter*, **18**, 7883-7892 (2006).

[17]K. D. Singh Mudher, M. Keskar, K. Krishnan, and V. Venugopal, Thermal and X-ray diffraction studies on Na_2MoO_4, $Na_2Mo_2O_7$ and $Na_2Mo_4O_{13}$, *Journal of Alloys and Compounds*, **396**, 275-279 (2005).
[18]V. L. Vinograd, D. Bosbach, B. Winkler, and J. D. Gale, Subsolidus phase relations in $Ca_2Mo_2O_8$-$NaEuMo_2O_8$-powellite solid solution predicted from static lattice energy calculations and Monte Carlo simulations, *Physical Chemistry Chemical Physics*, **10**, 3509-3518 (2008).
[19]E. Gürmen and E. Daniels, Crystal structure refinement of $SrMoO_4$, $SrWO_4$, $CaMoO_4$ and $BaWO_4$ by neutron diffraction, *Journal of Chemical Physics*, **55**, 1093 (1971).
[20]S. P. S. Porto and J. F. Scott, Raman spectra of $CaWO_4$, $SrWO_4$, $CaMoO_4$, and $SrMoO_4$, *Physical Review*, **157**, 716-719 (1967).

GLASS-CERAMIC STONEWARE AS A PROMISING APPLICATION FOR WASTE GLASSES

E. Bernardo♣,•
University of Padova, Dipartimento di Ingegneria Meccanica – Settore Materiali
via Marzolo, 9, 35131 Padova, Italy

L. Esposito, E. Rambaldi, A.Tucci
Centro Ceramico Bologna, Via Martelli 26, 40138 Bologna, Italy

ABSTRACT

One typical application of glasses from the melting of several inorganic waste is the manufacturing of glass-ceramics. These products generally exhibit interesting physical and mechanical properties, but their processing is quite complex and expensive. In this paper it is reported an alternative approach, consisting in replacing the feldspar flux, in the formulation of porcelain stoneware, with two waste glasses of the CaO-Al_2O_3-SiO_2 system, in the form of fine powders. For an optimized glass/clay ratio, the obtained 'glass-ceramic stoneware' samples, sintered at temperatures ≤1000°C, exhibited a complex of mechanical properties (bending strength approaching 90 MPa, fracture toughness exceeding 2 MPa $m^{0.5}$) similar to that of conventional porcelain stoneware, sintered at much higher temperatures. This was mainly attributable to the formation of several crystal phases upon sintering, due to both surface nucleation (yielding wollastonite or gehlenite) and chemical interactions between the glass and clay (yielding Ca-rich feldspars).

INTRODUCTION

Vitrification is a widely accepted technique for the stabilization of a number of inorganic waste, but it features a very significant drawback in the high energy consumption required by the melting processes.[1] A solution to this economic problem may result by considering the vitrification not as a final step, i.e. only a safe way to waste disposal, but as an intermediate point in the manufacturing of new materials, for which vitrified waste constitute basic raw materials. The benefits from selling the new materials from vitrified waste (i.e. from a "valorization" step) could compensate the costs of the vitrification process, with the additional economic and environmental advantage associated with the saving of natural raw materials.[1]

In a vitrification/valorization program we can consider a double strategy, aimed at optimizing both steps. First, costs and duration of glass-making may be drastically reduced by sudden cooling of the glass melt, with no refinement nor annealing, adopting a vitrification process resembling that of common ceramic frits; this solution could also provide an improved sealing of hazardous pollutants, since long treatments at high temperature (favorable to the volatilization of some oxides) are avoided. Second, the waste glass (in frit form) may be recycled in ceramic products for building applications, following different manufacturing routes. The most important is represented by the manufacturing of glass-ceramics, which could be considered a rather traditional field of application for waste glasses (a famous example is that of Russian "Slagsitalls", dating back to the 1960s):[2,3] the products are completely made of glass, so that there is a significant usage of waste glass for unit product, however it must be noted that glass-ceramics have a complex and rather expensive processing and also a specific and quite limited market.[1] Traditional ceramics could be an alternative, since waste glasses may be used as sintering aid for traditional formulations:[4-6] a limited quantity of glass is used for unit product,

♣ Corresponding author: enrico.bernardo@unipd.it
• Member, The American Ceramic Society

but traditional ceramics feature a mass market, very high production rates and also a very simple processing.

The present paper is intended to highlight a third way, consisting of the merging of the two valorization approaches reported above. Waste glasses are inserted in a significant amount in formulations for porcelain stoneware, in order to replace completely the feldspar fluxes (i.e. natural raw materials): the introduction of glass lowers the firing temperatures and also gives origin to new crystal phases.

EXPERIMENTAL

The chemical composition of the two glasses of the CaO-Al_2O_3-SiO_2 system considered in the present paper is reported in Table I. The first glass corresponds to the melting of a mixture of waste materials, as reported by Bernardo.[7] The second glass comes from the industrial melting of municipal solid waste incinerator fly ashes, performed by Europlasma (Bordeaux, France), by using plasma vitrification.[8] The two glasses were available in the form of large fragments, from the sudden cooling on cold steel substrates, subsequently ground to fine powders, with an average dimension (d_{50}) of 40 μm (determined by a laser granulometer).

Table I – Chemical composition of the starting materials (wt.%)

	G1 glass	*G2 glass*	*"C" Clay*	*"U" Clay*
SiO_2	61.5	40.7	80.8	64.8
Al_2O_3	12.2	12.6	13.8	21.9
CaO	12.7	32.2	0.1	0.5
MgO	0.8	3.3		0.5
Na_2O	5.3	4.6	0.1	0.5
K_2O	3.7	0.3	0.2	1.9
Fe_2O_3	2.0	1.0	0.1	0.9
Cr_2O_3		0.7		
F	0.5			
TiO_2	0.2	1.5	0.1	1.3
P_2O_5	0.1	2.2		
Other	0.8	0.9	5.0 (L.O.I.)	6.6 (L.O.I.)

The glasses were mixed with industrial clays, named C and U, normally used coupled to feldspar sands for the production of porcelain stoneware tiles. Both clays feature a similar content of kaolinite (about 40 wt.%), but while the rest of C clay is composed mainly by quartz, U clay contains quartz (about 40 wt.%) and other clay minerals, such as illite and smectite (about 15 wt.%), the rest being minor amounts of feldspar minerals. For G1 glass, it was considered the weight proportion of 60% glass/10% C clay/30% U clay, while for the G2 glass only U clay was employed, for a weight proportion of 70% glass/30% U clay. Aqueous slips were obtained by the addition of 33 wt.% distilled water and 1 wt.% deflocculant, followed by homogenization in a ball mill for 1 h. The slips were dried at 110°C overnight, sieved to granules with a dimension of about 100 μm and pressed in a cilindrical steel die (40 mm diameter), at 40 MPa. The obtained discs were subjected to thermal treatment, consisting in the heating at 10°C/min at different temperatures, with no soaking time and followed by natural cooling. Selected compositions were subjected to a faster treatment, at 40°C/min, with a 30 min soaking time at the firing temperature. The fired samples were subjected to mineralogical (qualitative and semi-quantitative X-ray diffraction via specific software packages),[9,10] microstructural (scanning electron microscopy, evaluation of the water absorption by immersion in water,[11] measurement of bulk density by the Archimedes' method, pycnometry) and mechanical analysis (evaluation of elastic modulus by non-destructive resonance frequency testing, measurement of strength from 4 point bending test applied on beam samples of about 3 mm × 2 mm × 35 mm - cut from larger samples and polished to a 3 μm finish, indentation testing).

RESULTS AND DISCUSSION

Fig.1a shows that the introduction of waste glasses causes a very substantial decrease of the temperature at which the water absorption is zero (1000°C for samples from G1 glass, 900-950°C for samples from G2 glass), compared to standard porcelain stoneware (1160°C), prepared from the same clays, but coupled to traditional feldspar sands. The reduction of the temperature for the sealing of the open porosity is consistent with the fact that glass softening occurs at a much lower temperature than those required for the melting of feldspars. Fig.1b shows that the Tg of the glass G1 is 640°C, while that of G2 is 710°C; the softening occurs consequently at about 700°C for G1 and 750°C for G2, while feldspars melt at temperatures in the order of 1100°C. The high content of glass modifiers, especially CaO, reasonably favors a low viscosity of the softened glasses, in turn favorable to a high densification.

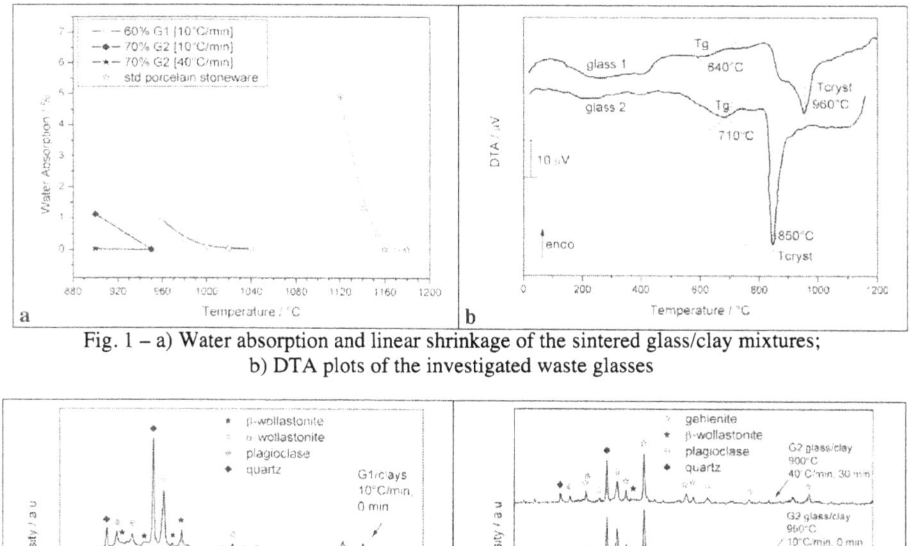

Fig. 1 – a) Water absorption and linear shrinkage of the sintered glass/clay mixtures; b) DTA plots of the investigated waste glasses

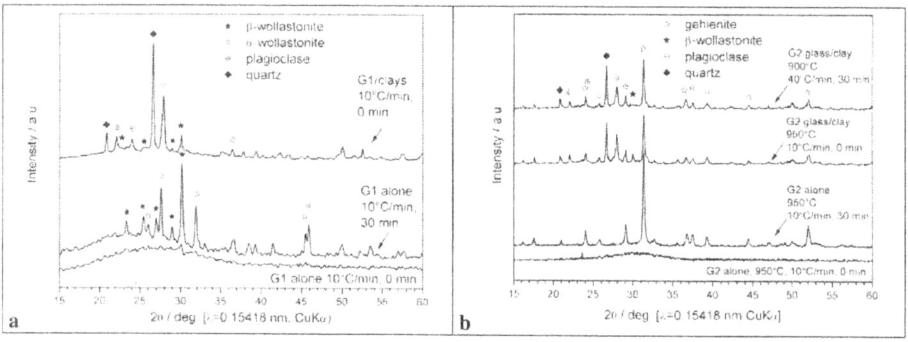

Fig.2 – Phase evolution in sintered glasses and glass/clay mixtures: a) G1 glass sintered at 1000°C; b) G2 glass at 900-950°C

Besides the sealing of the open porosity, also the shrinkage is particularly important. The shrinkage stabilizes around the temperatures required for zero water absorption: it is about 8% from

960 to 1040°C, for samples from G1 glass, and 6% between 900 and 950°C, for samples from G2 glass. This effect is consistent with the control of the viscous flow (preventing from viscous collapse) operated by crystal inclusions. In conventional porcelain stoneware the control of the viscous flow is due, besides to quartz residues, to mullite crystals, from thermal transformation of clay minerals. Consistent with the fact that mullite forms only in ceramics fired well above 1000°C, the ceramics from clay/waste glass mixtures do not show traces of mullite in the X-ray diffraction patterns. However, the mixtures exhibited alternative newly formed phases. These crystal phases are only partially attributable to the devitrification of the two glasses, which is expected since the firing temperatures are close to the crystallization peaks reported in Fig.1b (the peaks refer to fine glass powders; the peaks for coarser powders – not reported here for the sake of brevity - are less intense, the glasses being subjected to surface nucleation).[7,8] In fact, as illustrated by Fig.2, when pure fine glass powders are sintered, they yield only wollastonite, $CaSiO_3$ (in both α and β forms), and Na-Mg-Fe substituted gehlenite, $(Ca_{1.96}Na_{0.05})(Mg_{0.24}Al_{0.64}Fe_{0.12})(Si_{1.39}Al_{0.61}O_7)$, for G1 and G2 glass, respectively. The X-ray diffraction patterns for fired glass/clay mixtures, in the same Fig.2, feature additional phases, which can be all grouped in Ca-rich feldspars (plagioclases). These phases may develop from the interaction of metakaolinite, from the dehydration of clay minerals, with CaO and alkali oxides of the two glasses; the glass/clay interaction is consistent with the same "history" of clay minerals, which originated from the chemical weathering of feldspar minerals.[12] The reaction involving metakaolinite is a further reason for the observed absence of mullite crystals. Some kinetic effects must be noted: G1 and G2 alone crystallize only after 30 min at 1000°C and 950°C, respectively, while crystals appear after 0 min, at the same temperature, in glass/clay mixtures. The glass/clay interaction probably determines glass/crystal interfaces that promote further nucleation. Finally, it must be noted the presence of α form of wollastonite in G1 glass alone, and not in glass/clay mixtures, featuring plagioclase (Fig.2a), consistent with the findings about sintered glass-ceramics from G1 and pure kaolin clay added as inorganic binder.[7]

The ceramics here presented come from waste glasses in different amounts. This is due to the fact that with the same proportions of G1 glass, G2 glass gives ceramics with a very substantial porosity, higher than 10%, for all the conditions, with a not neglectible water absorption. This could be seen as a further consequence of the crystallization process associated with the sintering. Fig.1b shows that the crystallization peak of G2 glass is much more substantial than that of G1 glass: when activated, the crystallization of G2 glass likely proceeds very rapidly, and the increase of apparent viscosity provided by crystal inclusions[13] "freezes" the microstructural evolution in the G2 glass/clay mixtures, limiting the removal of internal pores. G1 glass, on the contrary, due to its less pronounced crystallization ability, is a good fluxing agent, even in limited quantity. For G2 glass, the sealing of open porosity corresponds only to the usage of 70% glass and only U clay: in this formulation the 10% of C clay is replaced by additional G2 glass. The residual porosity is reduced due to the increased amount of liquid phase, and also to the reduced the quantity of starting crystal inclusions, i.e. quartz residues, mostly present in C clay.

The results for samples with G2 glass are interesting in the light of the two types of treatment applied: the treatment with a high heating rate causes the water absorption to be zero at 900°C, 50°C below the level required by the treatment at low heating rate, previously applied also to samples with G1 glass. This is likely due to a reduced crystallization during the heating phase (due to the limited time at temperatures close to the crystallization peak), promoting the viscous flow; the crystallization is obtained during the soaking time. The advantageous effect of fast heating, in promoting the densification of G2 glass, undergoing rapid crystallization, has been recently reported for sintered glass-ceramics from the same glass, without clay additions.[8] Since the fast treatment is associated with a reduction of firing times of more than 40 min, substantial improvements of productivity are expected in prospect of large scale industrialization.

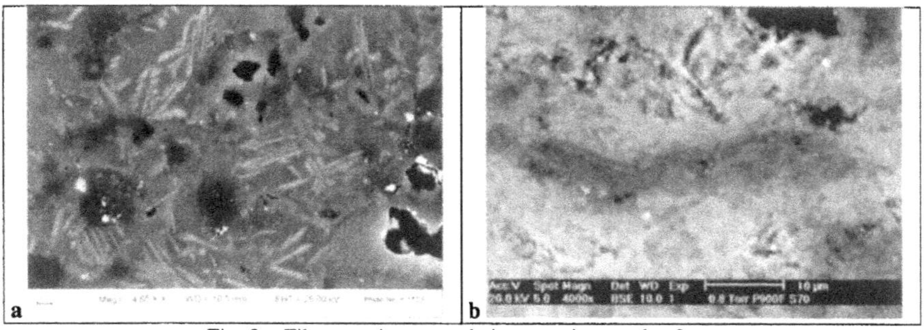

Fig. 3 – Fibrous microcrystals in ceramic samples from:
a) G1 glass, 1000°C; b) G2 glass, 900°C (fast heating)

From semi-quantititative X-ray diffraction analysis, the sample from G1 glass features about 62% glass, 12% residual quartz, 6% β-wollastonite and 20% plagioclase; the samples from G2 glass feature about 60% glass, 5% residual quartz, 13% gehlenite, 10% plagioclase, 4% β-wollastonite for both types of thermal treatment. Evidences of the crystallinity of the samples from both G1 and G2 glasses are reported in Fig.3.

Table II – Comparison of the mechanical properties of sintered glass/clay mixtures with those of standard porcelain stoneware

Sample type	Firing temperature (°C)	Total porosity (%)	Elastic modulus (GPa)	Bending strength (MPa)	Vickers' hardness (GPa)	Indentation K_{IC} (MPa $m^{0.5}$)
G2 [70%]/clay (40°C/min)	900	5.9	86.1 ± 1.9	69.6 ± 5.2	6.4 ± 0.2	
G2 [70%]/clay (10°C/min)	950	4.9	86.6 ± 0.8	71.5 ± 4.6		
G1 [60%]/clays (10°C/min)	1000	5.0	78.1 ± 1.8	88.3 ± 5.8	6.4 ± 0.3	2.2 ± 0.2
Standard porcelain stoneware	1160	3.0	77.1 ± 3.2	85.1 ± 7.1	6.5 ± 0.3	1.5 ± 0.1

Table II shows that the product of G1 glass/clay mixture is fully comparable to conventional porcelain stoneware: in addition to the porosity, the new product equals the traditional one also in elastic modulus, bending strength (about 90 MPa) and micro-hardness; the indentation fracture toughness (evaluated by using the equation of Anstis et al.)[14] is even improved, being about 2 MPa $m^{0.5}$. The color of the samples from G1 glass is light red: with a small addition of white pigment (i.e. zircon), the color is expected to become more similar to that of conventional porcelain.

Also the mechanical properties of the products from G2 glass/clay mixture are acceptable for most applications in the building industry; the bronze color achieved could be corrected by the application of glazes. It may be observed that the residual porosity of optimized samples from G1 and G2 glasses are comparable, but the properties are quite different. Samples from G2 glass possess a higher elastic modulus, but lower strength. In our opinion, this could be due to the different phase assemblage, and to the residual stresses from thermal expansion mismatch between the phases, which are still under investigation. Improvements in the properties of products from G2 glass are also being studied: a strategy is that of mixing the G2 glass with a secondary glass, aimed at modifying the balance of the three main oxides, CaO, Al_2O_3 and SiO_2, to values similar to those for G1 glass, in order to achieve a similar phase assemblage upon firing. This is potentially possible by the addition of a recycled glass rich in both alumina and silica, but poor in CaO, such as boro-alumino-silicate glass used in pharmaceutical applications.

CONCLUSION

Two waste glasses of the $CaO-Al_2O_3-SiO_2$ system have been successfully employed in the formulation of new clay based products, with mechanical properties suitable for application as tiles in the building industry, but sintered at much lower temperatures, i.e. ≤1000°C. The key point of the usage of waste glasses instead of traditional feldspar fluxes is, in addition to the possibility of significant viscous flow sintering at low temperatures, the crystallization of the same glasses, from both an "intrinsic" devitrification mechanism and a "extrinsic" glass/clay interaction, which replaces the contribution of mullite crystals in conventional clay based products.

REFERENCES

[1]P. Colombo, G. Brusatin, E. Bernardo, G. Scarinci, "Inertization and reuse of waste materials by vitrification and fabrication of glass-based products," *Curr. Opin. Solid State Mater. Sci.*, 7 (2003) 225-39.

[2]W. Höland, G. Beall, "Glass-ceramic technology," The American Ceramic Society (2002).

[3] R. D. Rawlings, J. P. Wu, A. R. Boccaccini, "Glass-ceramics: Their production from wastes- A Review," *J. Mater. Sci.*, 41 (2006) 733-61.

[4]G. Brusatin, E. Bernardo, F. Andreola, L. Barbieri, I. Lancellotti, "Reutilization of waste inert glass from the disposal of polluted dredging spoils by the obtainment of ceramic products for tiles applications," *J. Mater. Sci.*, 40 (2005) 5259-64.

[5]A. Tucci, L. Esposito, E. Rastelli, C. Palmonari, E. Rambaldi, "Use of soda-lime scrap-glass as a fluxing agent in a porcelain stoneware tile mix," *J. Eur. Ceram. Soc.*, 24 (2004) 83-92.

[6]Y. Pontikes, A. Christogerou, G. Angelopoulos, E. Rambaldi, L. Esposito, A. Tucci, "Use of soda-lime-silica scrap glass in the traditional ceramic industry," *Glass Technology*, 46 (2005) 200-7.

[7]E. Bernardo, "Fast sinter-crystallization of a glass from waste materials", *J. Non-Cryst. Sol.*, 354 (2008) 3486-90.

[8]E. Bernardo, E. Edme, U. Micron, N. Planty, "Fast-sintered gehlenite glass–ceramics from plasma-vitrified municipal solid waste incinerator fly ashes," *J. Am. Ceram. Soc.*, 92 (2009) 528-30.

[9]A.C. Larson, R.B. Von Dreele, "GSAS, General Structural Analysis System," Los Alamos National Laboratory Report, LAUR, 1994, 86-748.

[10] Match! program package (Crystal Impact GbR, Bonn, Germany), supported by data from PDF-2 database (ICDD -International Centre for Diffraction Data, Newtown Square, PA)

[11] ISO 10545-3 norm.

[12]W.E. Parham, "Formation of halloysite from feldspar: low temperature, artificial weathering versus natural weathering," *Clays and Clay Minerals*, 17 (1969) 13-22.

[13]A.R. Boccaccini, "On the viscosity of glass composites containing rigid inclusions." *Materials Letters*, 34 (1998) 285-8.

[14]G.R. Anstis, P. Chantikul, B.R. Lawn and D.B. Marshall, "A critical evaluation of indentation techniques for measuring fracture toughness: I, direct crack measurement," *J. Am. Ceram. Soc.*, 64 (1981) 533-8.

Solid Oxide Fuel Cells and Hydrogen Technology

LOW TEMPERATURE HYDROGEN RELEASE FROM BORONTETRAHYDRIDE-SODALITE AND ITS RELOADING: OBSERVATIONS IN IN-SITU AND EX-SITU TIR EXPERIMENTS

C. H. Rüscher*, F. Stemme, L. Schomborg, J.-Chr. Buhl

Institut für Mineralogie und ZFM, Leibniz Universität Hannover, Welfengarten 1, 30167 Hannover, Germany

*E-mail: c.ruescher@mineralogie.uni-hannover.de

ABSTRACT

Temperature dependent infrared (TIR) absorption experiments were used to investigate the mechanism of hydrogen release from borontetrahydride sodalite, $Na_8[AlSiO_4]_6(BH_4)_2$. The release of hydrogen is observed for temperatures between 150 and 450°C as indicated in situ by nitrate reduction. The hydrogen production follows the reaction $BH_4^- + 2\ H_2O \Rightarrow BO_2^- + 4\ H_2$ via consecutive intra cage reaction steps. Water molecules are injected from x-ray amorphous hydrosodalite species which are formed simultaneously with the borontetrahydride sodalite during synthesis. The hydrosodalite type species become reloaded with water under normal ambient conditions, which is used to proceed the hydrogen release. A first step for the reloading of hydrogen becomes obvious, too, when the reacted pellets where reheated under hydrogen flowing atmosphere.

INTRODUCTION

The synthesis of new material for hydrogen storage and their investigation of the release of hydrogen is of potential interest for future energy management. In this sense salts like $Al(BH_4)_3$, $LiBH_4$ or $NaBH_4$ are chemical compositions from which hydrogen can be obtained in destruction reactions with H_2O molecules. Since a breakthrough for the rapid reloading reaction with hydrogen using special designed catalytic metals has been reported such type of material indeed might substitute the present technologies in near future. It has also been shown that $NaBH_4$ can easily be used to form borontetrahydride sodalite, $Na_8[AlSiO_4]_6(BH_4)_2$, under mild hydrothermal conditions [1, 2]. This of course significantly reduces the possible hydrogen release per kg material $NaBH_4$. However, the handling of the material could be easier compared to the salts, i.e. the sodalite cage (Fig. 1) protects the BH_4^- group against moisture attack and also protects the environment. Additionally the BH_4^- destruction reaction could well be controlled on atomic level as the possible injection of water could be governed through the size of the sodalite windows which is a definite function of temperature [3]. Therefore it seems to be interesting to further investigate this new type of material for its possible potential in hydrogen release and its reload capabilities. Temperature dependent infrared (TIR) experiments using the KBr and NaCl pressed pellet technique have been reported to provide a powerful tool for the investigation of chemical reactions and structural phase transitions of sodalites and cancrinites [4]. These methods were used and further developed in order to investigate the mechanism of hydrogen release of borontetrahydride sodalite and its reloading. The main results are outlined here.

EXPERIMENTAL

The synthesis procedure of borontetrahydride-sodalite, $Na_8[AlSiO_4]_6(BH_4)_2$, followed that described by Buhl et al. [1] with 120°C temperature and 24 h reaction time using 0.5 g $NaBH_4$ salt (Merck 806373), 1g kaolinite (Fluka 60609) and 10 ml 16 M NaOH (Fluka 71691). The products were characterized using X-ray powder diffraction (Stoe Stadi P). Infrared spectra were taken using a FTIR spectrometer (Bruker IFS66v) with KBr and NaCl pressed pellets, a self constructed Ag sample holder (Fig. 2) and heating device as described in Ref. [4]. TG measurements (Setaram Setsys evolution 1650) were carried out under He flowing atmosphere (20 ml/min). Hydrogen reloading experiments were conducted using a furnace chamber heated at 400°C for 24 h with flowing hydrogen (50 ml/min).

RESULTS AND DISCUSSION

The obtained borontetrahydride sodalite, $Na_8[AlSiO_4]_6(BH_4)_2$, shows the same X-ray and infrared absorption characteristics as previously reported [1].The IR-absorption spectrum of this sample (Fig. 3) is characterized by the sodalite typical peaks of the framework vibrations at 450, 500, 666, 707, 734 and 1000 (all in cm^{-1}). The peaks at 1143 (ν_4), 2286 ($2*\nu_4$), 2241 (ν_3) and 2390 ($\nu_2+\nu_4$) can be assigned to vibrations of the BH_4^- tetrahedral group by direct comparison to the spectrum of pure sodium borontetrahydride salt. For the BH_4^- related absorption a strong broadening and decrease in peak height is observed with increasing temperature already at 150°C which could be related to the dynamical rotation due to a high degree of freedom. Indications of a new peak denoted as A in Fig. 3 appear, which increases in intensity and becomes clearly visible at 350°C and when the sample is cooled down again. The A feature indicates the first reaction step which leads to a release of hydrogen related with a small (less than about 5%) decrease in the intensity of the BH_4^- related peaks observed by comparison of the peak intensities before and after the TIR experiment up to 350°C. The borontetrahydydride sodalite sample shows a weak absorption feature at about 1650 cm^{-1} and a broad feature in the range between 3200 to 3700 cm^{-1} (D in Fig. 3) indicating the presence of H_2O species according to their bending and stretching modes, respectively. A sharper OH absorption feature is observed at about 3620 cm^{-1} (D' in Fig. 3). The broad H_2O related feature gradually decreases in intensity with increasing temperature up to 350°C and the sharper OH peak D' remains, which can be similarly observed in basic sodalites [5]. Cooling down again to 20°C shows that even the broader peak D still appears to some extend, indicating a remaining water content. It is essential that the loss in intensity of peak D is related to that water content consumed for the release of hydrogen for temperature up to 350°C and for the further heating up to 500°C (see below). The origin of features D and D' can thus be related to some X-ray amorphous (basic) hydro-sodalite type of material which is formed simultaneously with the borontetrahydride sodalite during synthesis. In the present case the X-ray amorphous part can be estimated to about 5 to 10 percent by weight. This interpretation has been worked out more recently increasing the amount of X-ray amorphous part and decreasing the borontetrahydride-sodalite contribution systematically [6]. In a combination of TG experiments with heating up to 500°C and characterisation by infrared absorption spectra it was also observed that the residual water content further reacts up to about 20 % of the total BH_4^- content (more details are given below) as described by the equation

$$BH_4^- + 2\,H_2O \Rightarrow BO_2^- + 4\,H_2 \qquad (1.)$$

and then stops. The sodalite containing the formed borate species remains stable at 500°C for the whole time investigated in so far (up to 100 h). However the reaction (1.) can be further continued reloading the hydrosodalite type part of the sample with water. This was done at 20°C under room ambient conditions as a typical characteristic of hydrosodalite. It was observed that in this way the total BH_4^- content was consumed in about 10 further runs of cooling down from 500°C, reloading with water and reheating and so forth.

Details of the formation of BO_2^- intra cage species and the nearly complete disappearance of BH_4^- absorption intensities is shown in another TIR experiment with a heating up to 600°C in Fig. 4. The BO_2^- absorption signature is denoted by C for the spectrum obtained after cooling to room temperature. The absorption feature denoted by B indicates another intermediate step which occurs within the reaction towards the final products as given in equation (1.). The intensity of A becomes reduces when B increases and B becomes reduced again when C increases showing the nature of consecutive reactions involved in eq. (1.). Moreover as discussed above the part of the hydro-sodalite type species with its water content can only account for a total reaction of not more that 20% of the borontetrahydride sodalite and that further reaction requires reloading of the hydrosodalite with water. Therefore the direct observation that the BH_4^- peak disappeared in the TIR experiment shown in Fig. 4 implies that the water molecules are gained from the (wet) NaCl matrix. The loss of water content from the NaCl plus sample pellet during heating can be seen by the appearance of "negative absorption" peak marked as HOH_{NaCl} in Fig. 4. On the other hand the peak marked by HOH_{Sod} can be assigned to the water content of the hydro-sodalite type contribution as a (wet) NaCl-pellet was used as reference that is why the effect of the water content corresponding to NaCl matrix is cancelled out in the spectrum taken at 20°C, but becomes negative for water loss larger than the content of HOH_{Sod}.

A further "negative absorption" signature is marked by arrows in Fig. 4. This effect can be explained by a reduction reaction induced by the hydrogen release of sodium nitrate which was supplied to the NaCl matrix by a special evaporation technique. A spectrum of sodium nitrate is given in Fig. 4 with a negative scaling for better comparison which proves the correct interpretation.

A first step for the reloading of hydrogen could be indicated observing the decrease in IR absorption intensity of BO_2^- species and increase in absorption intensities assigned to the borate species in the intermediate reaction steps, A and B (Fig. 5). For this reloading experiment the NaCl plus sample pellet effect as obtained after the 600°C TIR experiment was reheated under hydrogen flowing atmosphere (400°C, 24 h, 50ml/min). It seems to be interesting to investigate the reloading mechanism and the possibilities of further optimisation in future.

CONCLUSION

It has been shown that hydrogen is released from borontetrahydride sodalite using the presence of X-ray amorphous basic hydro sodalite species. The reaction occurs via intermediate reaction steps and stops when all the water content of the hydrosodalite species is consumed. The reaction proceeds reloading the water either through atmospheric contact (cooling down to room temperature and leaving the sample for short time at open conditions) or through an water containing matrix (NaCl). A first step of reloading hydrogen into the sodalite was observed.

[1] J.-Chr. Buhl, Th. M. Gesing, C. H. Rüscher, Microp. Mesop. Materials 80 (2005) 57-63.

[2] J.-Chr. Buhl, Th. M. Gesing, T. Höfs, C. H. Rüscher, J Sol. St. Chem. 179 (2006) 3877-3882.

[3] C. H. Rüscher, Th. M. Gesing, J.-Chr. Buhl, Z. Kristallographie 218 (2003) 1-10.

[4] C. H. Rüscher, Microp. Mesop. Materials 86 (2005) 58-68.

[5] G. Engelhardt, J. Felsche, P. Sieger, J. Am. Chem. Soc. 114 (1992) 1173.

[6] J.-Chr. Buhl, F. Stemme, C. H. Rüscher, submitted: Low temperature synthesis of BH4-enclathrated sodalite nano-particles from high alkaline solutions.

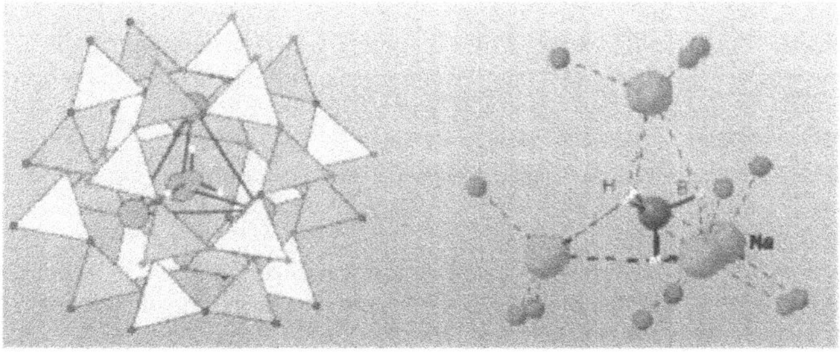

Fig. 1 AlO_4 and SiO_4 units forming the sodalite cage (left) and cage filling agent BH_4^- (right) with Na (pink): $Na_8(Al_6Si_6O_{24})(BH_4)_2$.

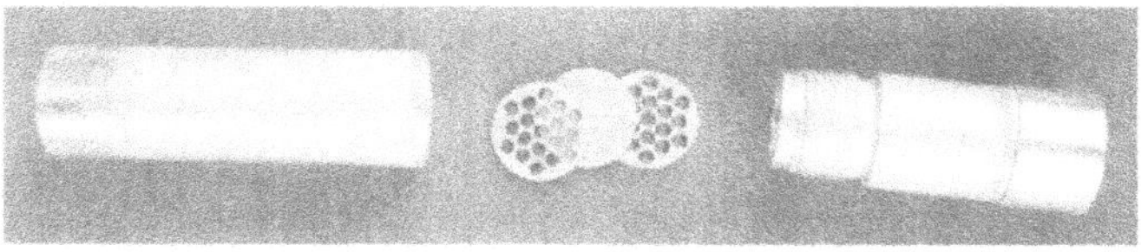

Fig. 2 Ag-tube, Ag net and KBr pellet as used in TIR experiments

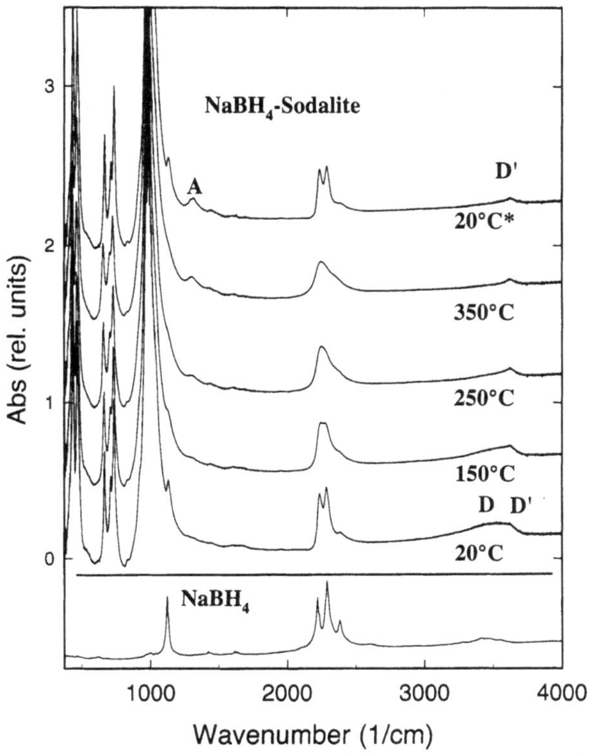

Fig. 3 Spectra of $NaBH_4$ and $NaBH_4$ enclathrated in sodalite at temperatures (in situ) as given. The 20°C* spectrum obtained after cooling down from 350°C compared to the initial ones shows the decrease of H_2O (D-peak) and BH_4^- and the formation of A type species („H_2BO") .

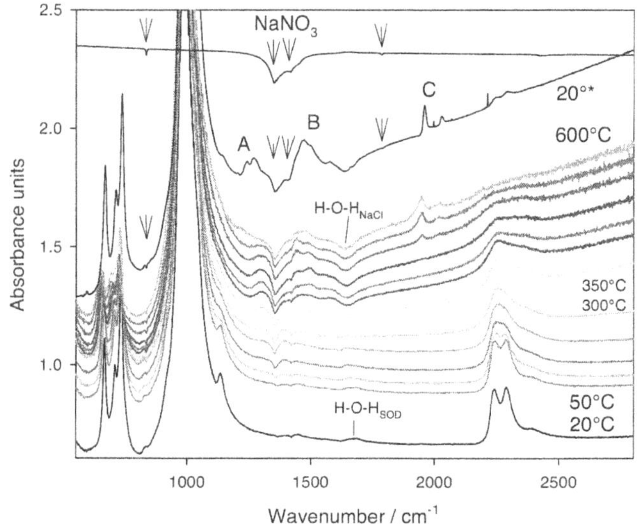

Fig. 4 TIR experiment in NaCl, in steps of 50°. 20°* denotes the spectrum after cooling with species A, B, C as in Fig. 3. Arrows mark the effect of tracer reaction of $NaNO_3 + H_2$.

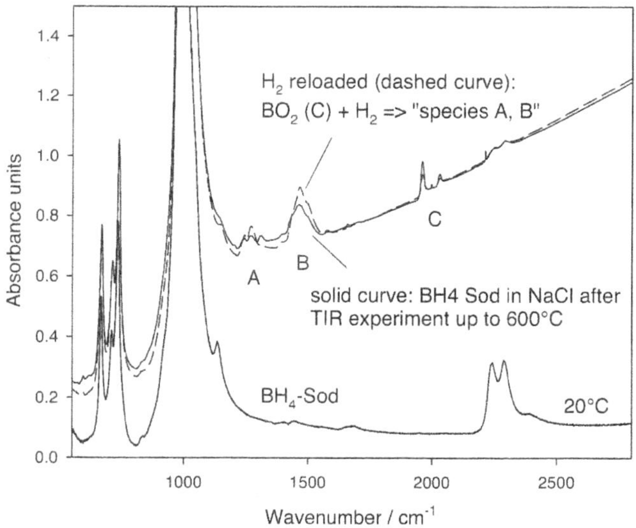

Fig. 5 Ex situ IR spectra: The H_2 reloading experiment using pellets

SYNTHESIS AND CHARACTERIZATION IN THE NEW SYSTEM $BI_2(AL/GA)_4O_9$ AND $^{18}O/^{16}O$ EXCHANGE EXPERIMENTS

T. Debnath[1], C. H. Rüscher[1], P. Fielitz[2], S. Ohmann[2], G. Borchardt[2]

[1]Institut für Mineralogie, Leibniz Universität Hannover, Callinstr. 3, 30167 Hannover, Germany
[2]Institut für Metallurgie, Arbeitsgruppe Thermochemie und Mikrokinetik, TU Clausthal, Robert-Koch-Straße 42, 38676 Clausthal-Zellerfeld, Germany

ABSTRACT

Compositions in the new system $Bi_2(Ga_xAl_{1-x})_4O_9$ (x = 0, 0.1, ..., 1) having mullite-type structure were synthesized using precursor gels obtained from the solution of corresponding nitrates in glycerin. The effects of Ga/Al substitution were characterized by IR absorption spectroscopy and X-ray powder diffraction (XRD). The linear change in lattice parameters related on x indicate ideal solid solution behaviour. The splitting of the high frequency vibration mode in the mixed composition indicate the presence of Ga-O_c-Ga, Ga-O_c-Al and Al-O_c-Al bonds, where O_c is the common oxygen of the double tetrahedral unit. $^{18}O/^{16}O$ isotope exchange experiments (at 800°C) show the extraordinary dominance of the O_c atom in this vibration. The relative absorption intensities of the M-$^{18}O_c$/M-$^{16}O_c$ vibrations used to estimate diffusion coefficient showing value of about $4*10^{-15}$ cm^2/s and $1.1*10^{-18}$ cm^2/s for $Bi_2Ga_4O_9$ and $Bi_2Al_4O_9$, respectively.

INTRODUCTION

Ceramics prepared by sintering using polycrystalline material in the system $Bi_{2-2x}Sr_{2x}Al_4O_{9-x}$ belong to the mullite type family $Bi_2(M'_{1-x}M_x)_4O_9$ [1] and show interesting conducting properties which could be related to the effect of oxygen diffusion via vacancies as required for applications in novel fuel cells [2 - 4]. However, recently our $^{18}O/^{16}O$ tracer diffusion experiments reveal diffusion coefficient between $8*10^{-18}$ and $8*10^{-16}$ cm^2/s [5] at 800°C ruling out extraordinary high oxygen diffusivity in these systems. Additionally, it has been shown that Sr can be only metastably substituted and at least partly segregates into an extra phase during heating procedures at 800°C. Within these investigations, the interpretation of the highest wavenumber infrared absorption peak as assigned to the M-O_c-M bond, i.e. with O_c the common oxygen of the characteristic double tetrahedral M_2O_7 units for this structure, becomes crucial. Therefore, in the present contribution the role of the O_c oxygen in the infrared absorption spectra was investigated and compared with existing oxygen diffusion data for this family of compounds. The new results are summarized here.

EXPERIMENTAL

The following chemicals (all Sigma Aldrich) were used: $Bi(NO_3)_3{\cdot}5H_2O$, $Al(NO_3)_3{\cdot}9H_2O$, $Ga(NO_3)_3{\cdot}7H_2O$, and $Fe(NO_3)_3{\cdot}9H_2O$, glycerine. The water content were checked by TG measurements (Setaram Setsys evolution 1650) and found to be as given with the chemical formula. Appropriate amounts of the nitrates in order to obtain nominal compositions $Bi_2(Ga_xAl_{1-x})_4O_9$, $Bi_2(Fe_xAl_{1-x})_4O_9$, and $Bi_2(Fe_xGa_{1-x})_4O_9$ (x = 0, 0.1,..., 1) were dissolved in minimum amount of glycerin. The homogeneous solutions was slowly heated at 80°C to form a viscous gel. The gel was heated at 120°C in an oven for about 12 h. During this process the gel swells into a fluffy mass. The dried gel was calcined in an open Pt crucible at 850°C for 48 h.

The products of nominal compositions $Bi_2(M_xM'_{1-x})_4O_9$ were characterized by X-ray powder patterns recorded in a Stoe Stadi P diffractometer (transmission geometry, $CoK\alpha_1$ radiation by a focusing Ge (111) monocromator, linear PSD). Lattice parameter refinements were performed using the Rietveld software Diffrac Plus TOPAS (Bruker AXS, Karlsruhe, Germany). For the calculation of the reflex profiles fundamental parameters were used on the basis of instrumental parameters calculated out of a silicon standard measurement. The zero point parameter was fitted and a polarization parameter was fixed.

Infrared absorption spectra were taken (FTIR spectrometer Bruker Vertex 80v) using PE and KBr pressed pellet technique in the spectral ranges 50 – 500 cm^{-1} and 370 – 4000 cm^{-1}, respectively. Using the overlapping range the spectra were merged with respect to the KBr range with 1 mg sample dispersed in 200 mg KBr. Spectra are given in absorbance units ($-lg(I/I_0)$, I_0, I are transmitted intensities through "reference" and "reference + sample", respectively). $^{18}O/^{16}O$ exchange experiments were carried out as reported earlier [5]. In particular preheated samples (96 h at 800°C under air conditions) were heated at 800°C for 16 hours in $^{18}O_2$ atmospheric conditions (95% enrichment and 200 mbar total pressure).

RESULTS AND DISCUSSION

The new series $Bi_2(Ga_xAl_{1-x})_4O_9$

It has been reported that the "Glycerine Method" reveals an easy way for preparing single phase polycrystalline materials for the system $Bi_2(Al_xFe_{1-x})_4O_9$ [6, 7] compared to other methods using glycine combustion technique [4] or standard solid state method, as was used to prepare the series $Bi_2(Ga_xFe_{1-x})_4O_9$ [8]. Thus, the "Glycerine Method" was used here to repeat the synthesis systematically for the two known systems and also to produce a new series of composition $Bi_2(Ga_xAl_{1-x})_4O_9$ with x = 0.0, 0.1, ..., 1.0. It was observed that all the diffraction peaks of all products could perfectly be indexed to the orthorhombic structure (space group = Pbam). Small amounts of Bi_2O_3 and Ga_2O_3, if present, could be taken away by washing procedures. The initial values of the lattice parameters used for the refinements were taken from Müller-Buschbaum et al. for $Bi_2Ga_4O_9$ [9] and from Abrahams et al. for $Bi_2Al_4O_9$ [3]. The refined lattice parameters for the end members of the new series well agree with the reported values. It is observed (Fig. 1) that the lattice parameters increase well approximates to linear relationship on x in all systems. The results of linear regression analysis are given in Tab. 1. Compared to earlier reports for the Fe/Al and Fe/Ga systems there is a significant less scattering in data, probably due to the improved synthesis technique.

The infrared absorption spectra (Fig. 2) of the samples of $Bi_2(Ga_xAl_{1-x})_4O_9$ solid-solution series are shown in the regions of 50-1100 cm^{-1}. Factor group analysis reveal for the compositions $Bi_2M_4O_9$ (space group Pbam) in total 33 IR active modes distributed according to 13 B_{3u} (E//a), 13 B_{2u} (E//b) and 7 B_{1u} (E//c) [10]. Using polarized single crystal reflectivity for $Bi_2Ga_4O_9$ spectra between 100 and 1200 cm^{-1} Beran et al. [10] could largely resolve 12, 12 and 6 peaks for the polarisations E//a, E//b and E//c, respectively. The missing three modes could be separated in the powder spectrum (Fig. 2) at 70 cm^{-1} (shoulder) and 75 cm^{-1} probably polarized within the 001 plane and at about 90 cm^{-1} (broad) attributed to E//c. In particular all infrared active modes superimpose in the powder spectra as claimed by Beran et al. [10] for $Bi_2Ga_4O_9$ and as shown by Rüscher [11] for mullite . However, any meaningful deconvolution of the powder spectra for the expected number of peaks shown in Fig. 2 between 150 and 800 cm^{-1} might be difficult or even impossible. In this sense the number of peaks observed by direct inspection might be obscured. As an example even the peak observed at highest wavenumber which has no component for E//c is composed of two with slightly different eigenfrequencies for E//a and E//b. The extraordinary high wavenumber is due to its for this structure type characteristic short M-O_c distance within the double tetrahedral unit M_2O_7 [10, 12]. According to Beran et al. [10] the direction of this bond deviates only slightly from the 110 direction. Therefore, the observed normal modes only slightly deviate in frequency (848 cm^{-1} for E//b; 852 cm^{-1} for E//a), which is not further separated in the powder related spectra. From inspection we obtain peak positions for Ga-O_c and Al-O_c of 850 and 922 cm^{-1}, respectively. The Ga-O_c distance (178.9 pm) is much shorter than the average distance <Ga-O> within the tetrahedral unit (184.3 pm). The same holds for $Bi_2Al_4O_9$, where Al-O_c is about 171.7 pm compared to the average <Al-O>, 177.3 pm [3]. According to this the observed splitting of the M-O_c peak in the mixed system must be related to the different M-O_c length of the Al_2O_7 and Ga_2O_7 double tetrahedral units. Consequently, a third M-O_c related peak as observed for the $Bi_2(Al_{1-x}Fe_x)_4O_9$ series [6] (compare Fig. 3a) implies a third type of cluster Al-O_c-Fe containing an averaging of the two shortest bonding length. Following this we also expect a third peak in the range of M-O_c vibrations of $Bi_2(Ga_xAl_{1-x})O_9$ mixed compositions. The overall shift of these peak related on x (Fig. 3 b) can be explained by the change in lattice parameters (Fig. 1), which may be seen for the lowest frequency modes in the spectra, too. The underlying splittings of all other modes due to details of the Ga/Al distribution may also be inferred from the loss of individual peak characteristics and the more envelope like appearance for the spectra of the mixed compositions (Fig. 2). The further analysis of this effect is however out of the scope of this work.

The $^{18}O/^{16}O$ isotope exchange experiments

An interesting observation has been the Al-$^{18}O_c$ extra peak observed in heat treated samples under $^{18}O_2$ atmospheric conditions of samples $Bi_{2-2x}Sr_{2x}Al_4O_{9-x}$ with probably small Sr concentrations $x < 0.025$ [5] This extra peak is shifted by about 40 cm^{-1} towards lower wavenumbers compared to the Al-$^{16}O_c$ related peak. Similar experiments has been repeated here for compositions $Bi_2(Ga_xAl_{1-x})_4O_9$ with x = 0.0, 0.5 and 1.0. For x = 0.0 and x = 1.0 an additional peak assigned to the isotope effect Al-$^{18}O_c$ and Ga-$^{18}O_c$ is observed as marked in the Fig. 4. For x = 0.5 the peak assigned to Ga-$^{18}O_c$ vibration is also well separated as marked. However, the other additional peaks, i.e. the Al-$^{18}O_c$ and Al-$^{16}O_c$-Ga related modes, are overlapped and may be obtained only in further fitting procedures. The pronounced isotope shift shows the extraordinary influence of the oxygen in this vibration, whereas all other absorption peaks observed in Fig. 4 are only slightly affected by the exchange. Thus, the O_c oxygen is not the only oxygen site involved in the oxygen diffusion. This may show a big potential in further investigation concerning specific sites for oxygen diffusion. Using such type of oxygen tracer exchange experiments a range of diffusion coefficients was estimated of about $8.7*10^{-18}$ to $8.7*10^{-16}$ cm^2/s for compositions in the system $Bi_{2-2x}Sr_{2x}Al_4O_{9-x}$ assuming average crystal sizes between 100 and 1000 nm and an exchanged surface layer of not more that about 5 volume percent related on the relative absorption intensities of Al-$^{18}O_c$ and Al-$^{16}O_c$ peaks [5]. From evaluation of our X-ray data we obtained an average crystal size between 400 and 600 nm for the $^{18}O_2$ treated samples. The obtained M-$^{18}O_c$/M-$^{16}O_c$ absorption intensities imply an exchange of not more than 5% in $Bi_2Al_4O_9$ and 30% in $Bi_2Ga_4O_9$. Assuming an average crystal size of 500 nm and an exchanged "surface layer of about z = 25 nm and 150 nm we may estimate (z = sqrt(D*t), t = 57600 s) diffusion coefficients of D = $1.1*10^{-18}$ cm^2/s and $0.4*10^{-14}$ cm^2/s for $Bi_2Al_4O_9$ and $Bi_2Ga_4O_9$, respectively. The value estimated for $Bi_2Ga_4O_9$ is in good agreement with the diffusion coefficient D = $2*10^{-14}$ cm^2/s obtained from single crystal isotope exchange experiments and SIMS depth profile measurement on a (010) plate [15]. The results are compared to data concerning oxygen tracer diffusion data [13] diffusion experiments on the formation and destruction of $(OH)^-$-groups in mullite [14] in Fig. 5, which were obtained at temperatures between 1200 and 1700°C. The data were extrapolated down to 800°C assuming an Arrhenius dependence of the diffusion coefficient. It can be seen that the oxygen diffusion estimated for compositions in the system $Bi_{2-2x}Sr_{2x}Al_4O_{9-x}$ is slightly enhanced compared to expected values for mullite and are still lower than for example proton related diffusion in mullite. The oxygen diffusion in $Bi_2Ga_4O_9$ significantly exceeds that of $Bi_2Al_4O_9$. Further experiments are on the way to check this more quantitatively.

CONCLUSION

The new series of composition $Bi_2(Ga_xAl_{1-x})_4O_9$ was prepared using amorphous gels obtained from a solution of metal nitrates in glycerine. According to lattice parameter variations related on x, this system shows ideal solid solution behaviour as observed for the systems $Bi_2(Al/Fe)_4O_9$ and $Bi_2(Ga/Fe)_4O_9$. The effect of varying Al/Ga contents was observed in the infrared spectra in particular by the distribution of absorption peaks assigned to the shortest M-O_c distances in the M_2O_7 double tetrahedral units. Oxygen isotope exchange experiments underline the extraordinary influence of the O_c oxygen within the infrared absorption spectra of these systems. The comparison of the oxygen diffusion data indicate that the interpretation of high oxygen diffusion for the observation of high conductivity [2-4] becomes problematic.

ACKNOWLEDGEMENTS

The authors are thankful for financial support by the DFG within PAK 279.

REFERENCES

[1] R. X. Fischer, H. Schneider, The mullite-type family of crystal structures. in: H. Schneider, S. Komarneni (Eds.), Mullite, Wiley-VCH, Weinheim, 2005, pp. 1–140.

[2] I. Bloom, M. C. Hash, J. P. Zebrowski, K. M. Myles, M. Krumpelt, Solid State Ionics 53-56 (1992) 739-747.

[3] I. Abrahams, A. J. Bush, G. E. Hawkes, T. Nunes, J. Solid State Chem. 147 (1999) 631-636.

[4] S. Zha, J. Cheng, Y. Liu, X. G. Liu, G. Y. Meng, Solid State Ionics 156 (2003) 197-201.

[5] C. H. Rüscher, Th. M. Gesing, T. Debnath, J.-Chr. Buhl, P. Fielitz, S. Ohmann, G. Borchardt, Z Krist Suppl. 29 (2009) 88.

[6] D. Voll, A. Beran, H. Schneider, Phys. Chem. Minerals 33(2006) 623-628.

[7] Th. M. Gesing, C. H. Rüscher, J.-Chr. Buhl, Z Krist Suppl. 29 (2009) 93.

[8] D.M. Giaquinta, G.C. Papaethymiou, W.M. Davis, H.-C. Zur Loye, J. Sol. State Chem. 99 (1992) 120-133.

[9] H. Müller-Buschbaum, D.C. de Beaulieu, Z. Naturforsch. B Chem. Sci. 33 (1978) 669-670.

[10] A. Beran, E. Libowitzky, M. Burianek, M. Mühlberg, C. Pecharroman, H. Schneider, Cryst. Res. Technol. 43(2008) 1230-1239.

[11] C.H. Rüscher, Phys. Chem. Minerals 23(1996) 50-55.

[12] C.H. Rüscher, Thermic transformation of sillimanite single crystals to 3 : 2 mullite plus melt: investigations by polarized IR-reflection micro spectroscopy, Mullite 2000, Elsevier Sci Ltd, Oban, Scotland, 2000, pp 2463-2469.

[13] P. Fielitz, G. Borchardt, M. Schmücker, H. Schneider, Philos. Mag. 87 (2007) 111-127.

[14] C. H. Rüscher, N. Eils, L. Robben, H. Schneider, J. Eur. Ceram. Soc. 28 (2008) 393-400.

[15] P. Fielitz, S. Ohmann, G. Borchardt, C. H. Rüscher, T. Debnath, unpublished result, it may also be noted that the single crystal plate were used to proof the existence of the Ga-^{18}Oc peak in polarized IR reflectivity.

Table 1 List of coefficients (m, p) of linear equation, $f(x) = mx + p$, obtained from the linear regression plot of the lattice parameters of $Bi_2(M_xM'_{1-x})_4O_9$ solid-solution series.

System	Axis	m	p
$Bi_2(Ga_xAl_{1-x})_4O_9$	*a*	0.209	7.7194
	b	0.184	8.1036
	c	0.202	5.6835
$Bi_2(Fe_xAl_{1-x})_4O_9$	*a*	0.048	7.9371
	b	0.143	8.3001
	c	0.109	5.8962
$Bi_2(Fe_xGa_{1-x})_4O_9$	*a*	0.253	7.7121
	b	0.325	8.1028
	c	0.311	5.6853

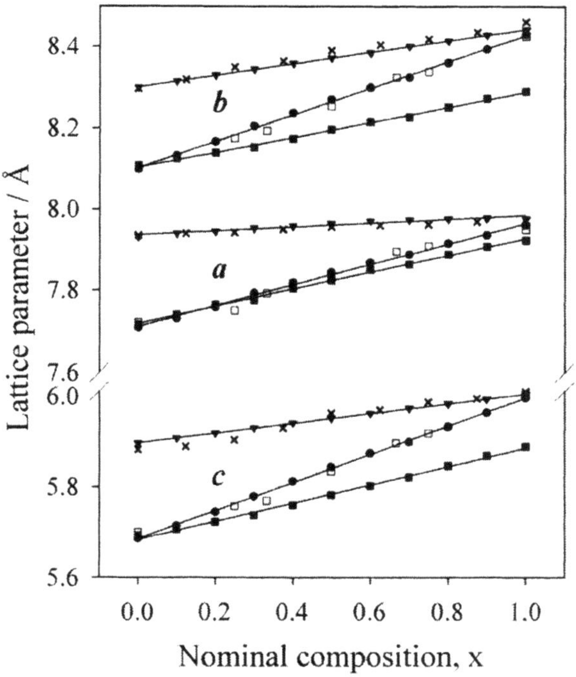

Fig.1 Lattice parameters for the solid solution series as a function of nominal composition x. The solid squares, circles and triangles represent the lattice parameters of the $Bi_2(Ga_xAl_{1-x})_4O_9$, $Bi_2(Fe_xAl_{1-x})_4O_9$, and $Bi_2(Ga_xAl_{1-x})_4O_9$ series, respectively, as synthesized in the present work. The solid lines represent results of linear regression of the lattice parameters. The lattice parameters of $Bi_2(Fe_xAl_{1-x})_4O_9$ (empty squares) from Voll et al. [6] and $Bi_2(Ga_xAl_{1-x})_4O_9$ (cross symbols) from Giaquinta et al. [8] are also shown.

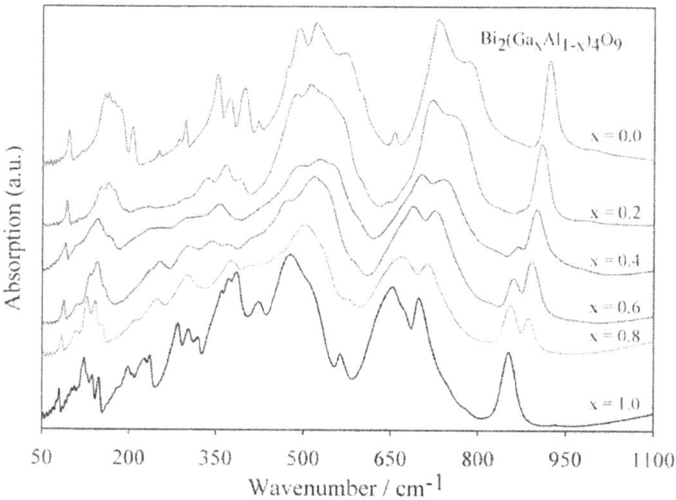

Fig. 2 FTIR absorption spectra of some selected samples of $Bi_2(Ga_xAl_{1-x})_4O_9$ solid solution series. The spectra are shifted vertically for the sake of clarity.

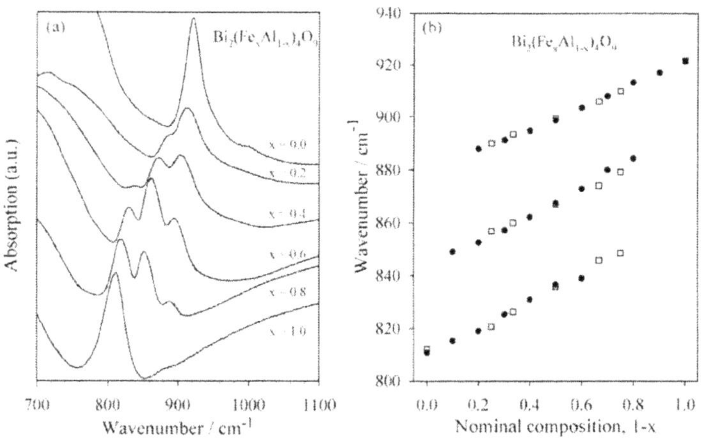

Fig. 3 (a) The three fold splitting of the high-energy mode assigned to M-O_c vibration in M_2O_7 unit of selected samples of $Bi_2(Fe_xAl_{1-x})_4O_9$. The spectra are shifted vertically for the sake of clarity. (b) Positions of peak maxima as a function of x. The band positions (symbols assigned by empty squares) reported by Voll et al. [6] are also given.

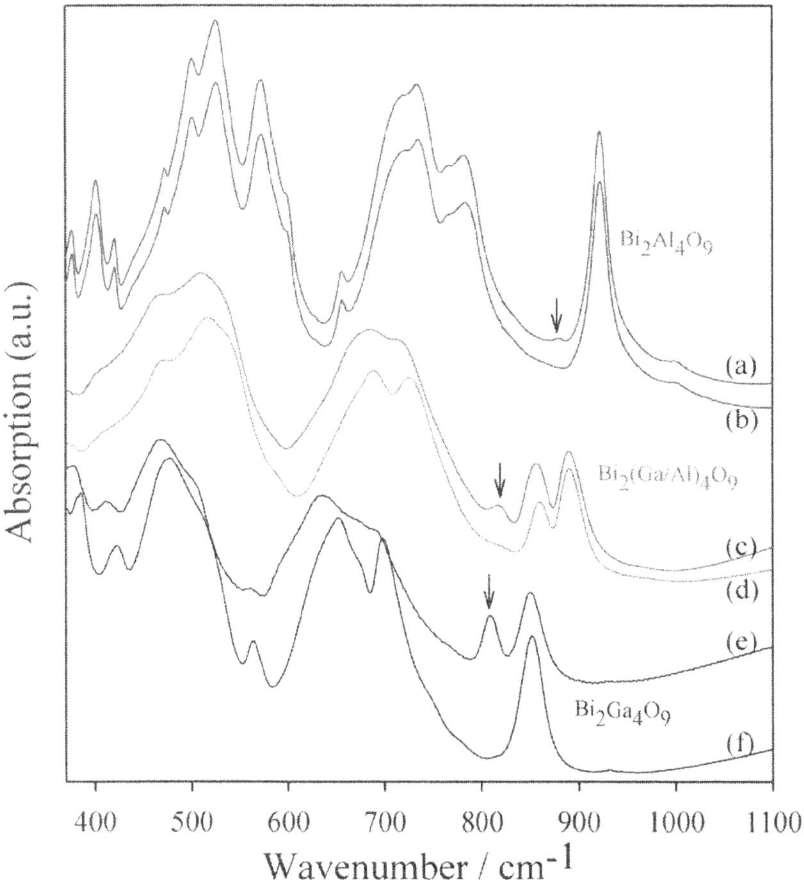

Fig. 4 FTIR absorption spectra of samples with compositions $Bi_2Al_4O_9$ (a, b), $Bi_2(Ga_{0.6}Al_{0.4})_4O_9$ (c, d) and $Bi_2Ga_4O_9$ (e, f) of in air heated (800°C/96h) before (a, c, e) and after (b, d, f) heating in $^{18}O_2$ (800°C/16h).

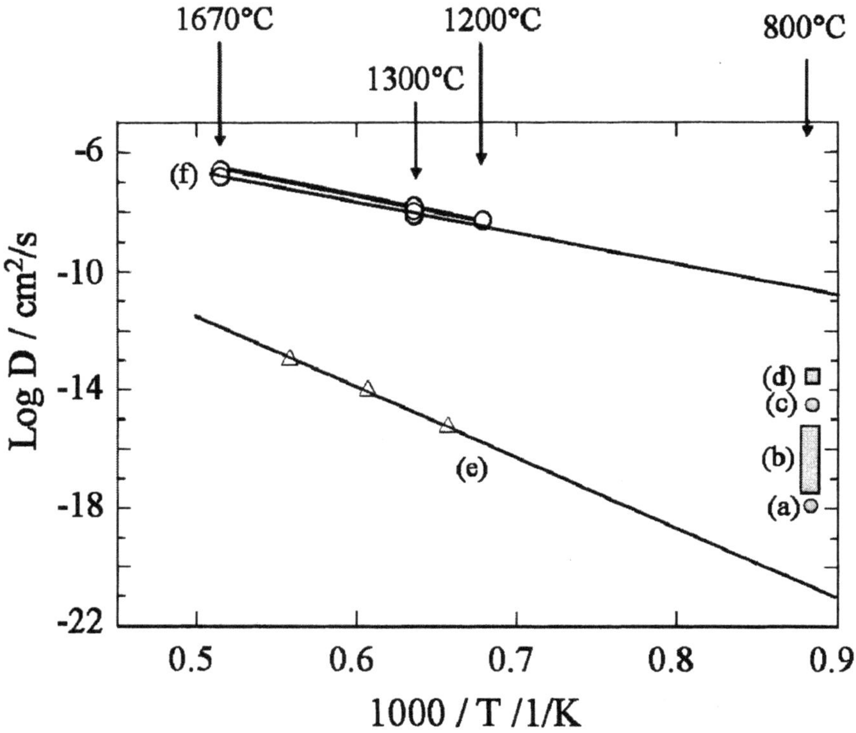

Fig. 5 Estimated diffusion coefficients from $^{18}O/^{16}O$ tracer diffusion experiments using M-Oc infrared absorption intensities data: (a) $Bi_2Al_4O_9$ and (b) $Bi_2Ga_4O_9$ (this work); (c) $Bi_{2-2x}Sr_{2x}Al_4O_{9-x}$ [Ref. 5]. Calculated diffusion coefficients from single crystal data: (d) $Bi_2Ga_4O_9$ single crystal data [Ref. 15]; (e) single crystal data of 2:1 mullite [Ref. 13]; (f) obtained for the formation and destruction of OH groups in single crystal mullite [14]. Solid line due to linear regression.

MICROSTRUCTURE OF Sm-DOPED CERIA INTERLAYER AND THE IMPEDANCE OF $LaSrCoFeO_3$/SDC/LAMOX HALF CELL

Yu-Chen Chen, Hun-Chieh Chang, Dah-Shyang Tsai*, Wen-Hung Chung , Minh-Vien Le

Department of Chemical Engineering, National Taiwan University of Science and Technology, 43, Keelung Road, Taipei 10607 Taiwan

* Corresponding author

E-mail address: dstsai@mail.ntust.edu.tw

ABSTRACT

A thin interlayer of samarium doped ceria (SDC) is applied as the diffusion barrier between $La_{1-x}Sr_xCo_yFe_{1-y}O_3$ x=0.1 - 0.4, y=0.2 - 0.8 (LSCF) cathode and $La_{1.8}Dy_{0.2}Mo_{1.6}W_{0.4}O_9$ (LDMW82) electrolyte, to obstruct Mo-Sr diffusion and solid state reaction. We demonstrate the effectiveness of diffusion barrier in contrasting the clearly defined interfaces of LSCF/SDC/LDMW82 against a rugged growing product layer of LSCF/LDMW82 at 800 °C, and analyzing the product composition and the probable new phase. Additionally, the measured polarization resistance is considerably lower for the half cell with a diffusion barrier. The electrochemical performance of LSCF cathode is investigated on a half cell built on the SDC protected LDMW82. The half cell of LSCF x=0.4 persistently outperforms that x=0.2 in polarization resistance, because of its small low-frequency contribution. The activation energy of polarization resistance is also lower for $La_{0.6}Sr_{0.4}Co_yFe_{1-y}O_3$, 112 - 135 kJ/mol, while that for $La_{0.8}Sr_{0.2}Co_yFe_{1-y}O_3$ is 156 - 164 kJ/mol. $La_{0.6}Sr_{0.4}Co_yFe_{1-y}O_3$ y=0.4 - 0.8 is the proper composition for the cathode interfaced to SDC/LDMW82. A preliminary result on the single chamber SOFC of LSCF6482/SDC/LDMW82/GDC+Ni shows a maximum power density of 126 mW cm^{-2} at 600 °C in the mixture of methane and air.

INTRODUCTION

ABO_3 oxides of perovskite structure have been the main candidates in the search for a proper cathode composition for solid oxide fuel cell (SOFC), owing to their low polarization losses and less materials cost. An early study on matching yttria-stabilized zirconia (YSZ) with $(La,Sr)CoO_3$ and other oxides of mixed conductor revealed the importance of diffusion and reaction between cathode and electrolyte [1]. The subsequent studies demonstrated that the components of La_2O_3 and SrO in perovskites might react with the zirconia component in YSZ and produce undesired $La_2Zr_2O_7$ and $SrZrO_3$. Mn dissolution could also occur at the interface of $(La,Sr)MnO_3$ and YSZ at high temperatures [2,3].

LAMOX denotes a family of oxygen ion conductors based on its parent crystal $La_2Mo_2O_9$, whose β polymorph is much more conductive than YSZ [4-6]. High-valence molybdenum is the key element for such a high ionic conductivity. Meanwhile the compatibility problem between electrodes and LAMOX is also centered on Mo [7]. Researchers have reported difficulties in finding an appropriate cathode composition for the LAMOX electrolyte, especially because many excellent cathodes contain strontium which reacts with molybdenum and produces $SrMoO_4$ [7,8]. To avoid this problem, we have synthesized the porous cathodes of lanthanum calcium cobalt ferrite (LCCF), and evaluated their performances in direct contact with LAMOX. The minimum polarization resistance of LCCF at 700 °C has been measured 14 - 18 Ω cm^2 using impedance spectroscopy [9,10]. The polarization resistance may be further reduced by a more conductive perovskite cathode, such as lanthanum strontium cobalt ferrite (LSCF), if diffusion and reaction between LAMOX and LSCF can be obstructed. Compared with YSZ, ceria exhibits less reactivity with the perovskite materials. Gadolinium doped ceria (GDC) has been reported physicochemically compatible with $La_2Mo_2O_9$ up to 1000 °C [11]. The ceria-based interlayer has also been adopted as a diffusion barrier to prevent the interfacial reaction between LSCF and YSZ [12-15].

In this work, we apply a layer of samarium doped CeO_2 (SDC) on the electrolyte of lanthanum dysprosium tungsten molybdate (LDMW) to obstruct its probable reaction with the LSCF cathode. Negative aspects of the Mo migration between LDMW and LSCF are made evident when comparing microstructures and impedances of two cells with and without SDC. The validity of SDC on LAMOX, the LSCF performance is evaluated on the SDC protected LDMW over an extensive compositional range in a half cell. Furthermore, we report a test-run result on a single chamber SOFC based on the LSCF/SDC/LAMOX anode structure.

EXPERIMENTAL

Preparation of pastes and electrolyte

The SDC paste of $Ce_{0.8}Sm_{0.2}O_{1.9}$ was prepared via the citrate gel route, which began with dissolving the stoichiometric amounts of $Ce(NO_3)_3{\bullet}6H_2O$ and $Sm(NO_3)_3{\bullet}6H_2O$ in deionized water, then mixing with citric acid to form a precursor solution. The solution was prepared with a molar ratio of metal cations : citric acid = 1 : 2, and the rare-earth concentration was 0.5 M. This chelated solution was heated at 70 °C in a water bath to thicken, and became a gel after ~4 h. The semi-rigid gel was further decomposed into a foamy powder at 140 °C. The foamy powder was pulverized and calcined at 800 °C for 3 h in air. Aggregates of the calcined powder were crushed in ball milling, using ethyl alcohol and zirconia balls. The dried SDC powder was ball milled again with α-terpineol and ethyl cellulose to prepare a paste for screen printing. The final solid content of SDC paste was ~30 wt%.

The LSCF paste for screen printing was calcined at 1000 °C for 2 h. The paste and its derived cathode are denoted according to the cation composition. For example, $La_{0.6}Sr_{0.4}Fe_{0.8}Co_{0.2}O_3$ is designated LSCF6482, with the first two digits for A-site composition and the other two digits for B-site composition. The electrolyte $La_{1.8}Dy_{0.2}Mo_{1.6}W_{0.4}O_9$ is designated LDMW82, because of 80% Mo and 20% W in the molybdenum site. LDMW82 was prepared via the solid state reaction route, using high purity chemicals of La_2O_3, Dy_2O_3, MoO_3, WO_3. The powder mixture was calcined at 880 °C, following by pulverization and milling with proper organic binders. The relative density of the electrolyte disk was higher than 95% after sintering at 950 °C for 6 h.

Impedance analysis and the half-cell and the single chamber cell measurements

The half cell used in impedance measurement adopted the symmetric electrode configuration. In fabricating the half cell, a LDMW82 disk of thickness 1.2 mm was polished on both sides and screen printed with the SDC paste. The SDC protected disk appeared glossy after firing at 1000 °C for 2 h. The LSCF paste was screen printed on both sides of SDC/LDMW82/SDC in an aligned position, and fired at 800 °C for 2 h. The porous LSCF electrode was 9 mm in diameter and ~10 μm in thickness. Two porous Pt layers served as current collectors, prepared by screen printing the Pt paste (C3605, Heraeus, Germany) on the LSCF electrodes and firing at 850 °C.

The half cell was placed in a tubular furnace with an inserted thermocouple for temperature measurement. Its impedance spectrum was recorded using a Solartron 1260 frequency analyzer. The impedance data were collected at zero dc bias, in the frequency range 0.01 Hz – 1.0 MHz at the perturbation amplitude 50 mV. All measurements were conducted in air, with O_2 partial pressure 0.21 atm.

The single chamber cell was prepared based on the configuration of LSCF6482/SDC/LAMOX/ GDC+NiO, and the current collectors were Ag instead of Pt. I-V data of the single chamber cell were collected in flowing methane/air with the CH_4/O_2 ratio equal to 1. The cell was reduced first, following by raising the furnace temperature to the desired temperature.

Structure characterization

The phases of sintered samples or powdery samples were examined in X-ray diffraction experiments, using a diffractometer equipped with a Cu Kα radiation source (D/Max-RC, Rigaku, Japan). The average crystallite size was estimated by the broadened peak width using the Scherrer formula, $0.94\lambda/(B\cos\theta)$, in which λ was the wavelength and B was the full width at half-maximum at the Bragg angle θ. Morphological features of the diffusion barrier layer and the product layer between LSCF and LDMW were analyzed using a field-emission scanning electron microscope (JSM6500F, JEOL, Japan).

RESULTS AND DISCUSSION

SDC interlayer

A reaction product layer will grow at the SOFC operation temperature when the two are in direct contact. We annealed the interface of LSCF6482/LDMW82 at 800 °C for 6 h, so that the product layer at the interface grew thick enough for EDS analysis. After the thermal annealing, the product layer became ~1.1 μm in thickness, as shown in Figure 1(b). The elemental analysis reveals that the product layer contains 13 mol% Mo, 7 mol% Sr, 9 mol% La, 2 mol% Co, 1 mol% W, and the balance is oxygen. The characteristic X-ray signals for Fe and Dy are indistinguishable from the background noise. Since La is the element common to LDMW and LSCF, the result suggests the main diffusing cations are Mo and Sr. Away from the product layer, the EDS analysis also detects the presence of Mo in the coarsened LSCF6482 layer, ~2 mol%. Examining this micrograph indicates Mo migration into LSCF6482 leads to considerable grain coarsening above the product layer, and strange protrusions underneath the product layer.

In contrast, there is no product layer at the interface of LSCF6482/SDC/LDMW82, as shown in Figure 1(a), because a porous SDC layer of 0.6 μm effectively impedes the cation diffusion under the same thermal annealing conditions. The Mo signal of LSCF6482, lying above the SDC layer, is analogous to the background noise. Underneath the SDC layer, the interfacing LDMW82 is flat and clearly defined. Still, it is reasonable to speculate on probable diffusion between SDC and LDMW82, since the sintering temperature of SDC interlayer was quite high, 1000 °C. In a separate experiment, a thick SDC layer of ~24 μm was printed on LDMW82 and sintered at 1000 °C for 2 h. No product layer was found between SDC and LDMW82. EDS analysis of the thick SDC, near the interface, indicates 19 mol% Ce, 3 mol% Sm, 3 mol% La, 3.5 mol% Mo. This result points out that Mo and La did diffuse into SDC, but only a small amount. We conclude that LDMW and SDC crystals can tolerate a small amount of cation interdiffusion without forming a new phase to disrupt the interface structure. To check the effectiveness of SDC diffusion barrier for every LSCF composition of this work, we examined every LSCF/SDC/LDMW82 interface and assured it was clear and defined using electron microscopy.

The probable new phase involved in the solid state reaction between LDMW82 and LSCF6482 is shown in Figure 2, which illustrates the XRD pattern of their powder mixture annealed at 800 °C for 6 h. Also plotted are XRD patterns of the calcined LDMW82 and the calcined LSCF6482. In addition to the diffraction lines of LDMW82 and LSCF6482 of reactants, there are several new lines in the pattern of fired LDMW82+LSCF6482 mixture. The new peaks at 27.2, 29.4, 33.1° are attributed to $SrMoO_4$, those at 28.4, 30.7, 36.6, 44.1° could belong to Sr_3MoO_6 [16]. The unidentified phase might be a Mo-deficient lanthanum molybdate. The presence of $SrMoO_4$ crystal has been reported in the reaction between (La,Sr)MnO_3 and $La_2Mo_2O_9$ [7,8].

The electrochemical impedance spectroscopy provides a more sensitive measure for detecting the product layer between LDMW and LSCF. Without the extra thermal annealing, the impedance of LSCF6482/LDMW82 cell is substantially larger than that of LSCF6482/SDC/LDMW82 cell. Figure 3 compares 700 °C impedance spectra of these two half cells, which were fabricated using the same batches of LSCF6482 and LDMW82. The impedance arc of the LSCF6482/SDC/LDMW82 cell is located at the left-hand side of the larger arc of LSCF6482/LDMW82, indicating that the negative influence on impedance is evident even without prolonged 800 °C annealing. Despite the apparent size difference, the shapes of two arcs are quite similar. The two arcs are elongated, suggesting the reaction product or the SDC interlayer contributes to the polarization resistance. An equivalent circuit, illustrated in Figure 3, is applied to fit the impedance data. The equivalent circuit includes (R_1/CPE_1)

in series with GE_1 to account for the high-frequency contribution and the low-frequency contribution of oxygen reduction reaction (ORR) in the impedance arc. GE_1 is the Gerischer-type element that results from the slow processes of oxygen ion diffusion and surface reaction in a semi-infinite porous cathode. Following the Adler's notation [17-19], we express the complex impedance of GE_1 as $R_{chem}\sqrt{\frac{1}{1+jt_{chem}\omega}}$, in which R_{chem} and t_{chem} are characteristic resistance and time constant for these non-charge-transfer processes, j the imaginary number, ω the frequency. The resistance R_1 and the constant phase element CPE_I are used to describe the impedance contributions of interface, including the thin SDC interlayer (or the product layer), the electron-transfer, the ion-transfer processes at the interfaces [17-20]. Fitting the two arcs of Figure 3, with the proposed equivalent circuit, yields the R_1 value 21.9 Ω for the cell without SDC interlayer, much larger than 1.9 Ω for the cell with SDC diffusion barrier. The R_{chem} value without SDC is also larger 4.24 Ω, compared with that with SDC 0.47 Ω. The electrolyte resistance R_S seems to be the least affected, 10.1 Ω for the cell without SDC is still higher than 6.7 Ω for the cell with SDC. The above example suggests that the solid-state reaction between LDMW and LSCF affects every component of the cell impedance.

LSCF solid solutions

The XRD patterns of calcined LSCF powders exhibit the structural features of perovskite, no secondary phase has been detected. Four typical diffraction patterns are illustrated in Figure 4 to show the influences of Co content, while the other four diffraction patterns in Figure 6 depict the influences of Sr content. The diffraction lines are indexed on a rhombohedral cell with space group $R\bar{3}c$, although the patterns of low Sr and Co contents may be more adequately indexed on an orthorhombic cell. Figure 4 demonstrates the peak positions of LSCF with 40% Sr shift toward higher angles with increasing Co content, signifying the lattice shrinks as smaller Co^{3+} substitutes for Fe^{3+} in the B site [22,23]. The other LSCF series with 20% Sr display the same trend. On the other hand, the diffraction lines of LSCF with 80% Co barely shift with increasing Sr content, as shown in Figure 5. The nearly constant cell volume may be explained by two counteractive effects of Sr substitution. The expansion effect of larger Sr^{2+} replacing La^{3+} is compensated by the shrinkage effect of an increase in the average valence of B-site cations, which is necessary for the charge neutrality in introducing acceptor solute Sr^{2+} in the A site. In this work, we focus on the performance of LSCF containing 20 or 40% Sr and various amount of Co. The electrical conductivity of $La_{1-x}Sr_xCo_{0.2}Fe_{0.8}O_3$ x=0.2-0.4 has been reported at their maxima at 550 - 800 °C [23], and the maximum conductivity of $La_{0.8}Sr_{0.2}Co_yFe_{1-y}O_3$ y=0.2-0.8 has been recorded at 550 - 900 °C [22].

The crystallite size of LSCF calcined powder varies slightly from batch to batch for different compositions. The LSCF crystallite size, estimated using the line width of the (024) reflection, varies between 15 and 35 nm. The crystallite size of SDC is estimated ~22 nm, using the line width of the (111) reflection. These oxide powders in the pastes are of high sinterability, which is an essential attribute for firing with the less refractory LAMOX electrolyte.

Cathode performance of LSCF/SDC half cell

Since a thin SDC interlayer has been demonstrated as an effective diffusion barrier and the conductivity of LSCF is high in the SOFC operation temperature range, the following cathode performance is evaluated using the cells with an SDC protected LDMW electrolyte. Figure 6(a) compares the impedance spectra of $La_{0.6}Sr_{0.4}Co_yFe_{1-y}O_3$ (y=0.2 - 0.8) at 700 °C. The impedance arc shrinks rapidly with increasing cobalt content. Meanwhile the frequency at maximum –Im(Z) value increases significantly with increasing cobalt content. The equivalent circuit analysis reveals the cobalt influences in more details. Table I lists the values of time constant t_{chem} and the area-specific values of R_{chem} and (R_{chem}+R_1). These characteristic values decrease with increasing cobalt content, consistent with the literature reporting the electronic and ion conductivities of LSCF increase with increasing Co content [22], and the high ion diffusivities and oxygen vacancy concentrations of cobaltite perovskites improve their electrochemical performance [24].

On the other hand, the impedance arc size of $La_{0.8}Sr_{0.2}Co_yFe_{1-y}O_3$ is generally larger than that of $La_{0.6}Sr_{0.4}Co_yFe_{1-y}O_3$. Figure 6(b) shows the impedance arc of $La_{0.8}Sr_{0.2}Co_yFe_{1-y}O_3$ (y=0.2 - 0.8) at 700 °C shrinks with increasing Co content at a slower pace. The fitted values of t_{chem}, R_{chem}, and (R_1+R_{chem}) for $La_{0.8}Sr_{0.2}Co_yFe_{1-y}O_3$ are also listed in Table I, in which the fitted resistance data are converted into area-specific values by multiplying with the electrode area and divided by two, due to the symmetry of electrodes.

A low value of polarization resistance (R_1+R_{chem}) is desired, since it leads to a low polarization loss during SOFC operation. Among the LSCF compositions we have studied, LSCF6482 seems to be the best cathode composition. The area-specific value of (R_1+R_{chem}) is 0.77 Ωcm^2 at 700°C, and 0.45 Ωcm^2 at 800°C. These polarization resistance values appear to be slightly larger than the polarization resistances of LSCF reported in the literature. The impedance resistance of $La_{0.8}Sr_{0.2}Co_{0.8}Fe_{0.2}O_3$/SDC was reported 0.3 Ωcm^2 at 700 °C [25], that of $La_{0.6}Sr_{0.4}Co_{0.95}Fe_{0.05}O_3$/SDC was 0.13 Ωcm^2 at 700 °C [26], that of $La_{0.8}Sr_{0.2}$ $Co_{0.5}Fe_{0.5}O_3$/YSZ was 0.67 Ωcm^2 at 680 °C [27], that of A-site deficient $La_{0.78}Sr_{0.2}Co_{0.2}Fe_{0.8}O_3$ on CGO/YSZ composite electrolyte was 0.4 Ωcm^2 at 850 °C [28]. The higher polarization resistance of this work is attributed to the interfacial resistance R_1, because of the SDC layer involvement.

Figure 7 presents the typical variation in LSCF impedance with respect to temperature. The arc size quickly decreases with increasing temperature and the frequency at maximum –Im(Z) value increases significantly with increasing temperature. The entire impedance curve of LSCF6428 moves to the left as the electrolyte resistance retreats toward a lower value.

Figures 8 and 9 summarize the cathode performance of all LSCF compositions in Arrhenius plots. Figure 8 presents the temperature dependence of R_{chem} and (R_1+R_{chem}). There are two groups of Arrhenius plots for R_{chem} in Figure 8(a), which shows the R_{chem} data of LSCF with 20% Sr (LSCF82) are consistently above those of LSCF with 40% Sr (LSCF64), except LSCF6428. The correlation lines for LSCF82 also show stronger temperature dependence. The activation energy for LSCF82 ranges from 137 to 174 kJ/mol. While the activation energy for LSCF64 is quite similar, 134 - 144 kJ/mol. The R_{chem} values for LSCF6482, 6464, and 6446 are evidently less than those of the rest compositions.

The Arrhenius plots of (R_1+R_{chem}) are shown in Figure 8(b). The variation of (R_1+R_{chem}) with respect to the LSCF composition is less systematic, because R_1 is involved with the troublesome but essential SDC layer. Nonetheless, the temperature dependence of (R_1+R_{chem}) still resembles Figure 9(a). And the three compositions of LSCF6482, 6464, 6446 exhibit an improved cathode performance over the previously studied LCCF. Their (R_1+R_{chem}) values at 700 °C are a factor of 10 less than the polarization resistance of LCCF/LDMW82 (14 - 18 Ω cm^2) [9,10]. The temperature dependence of (R_1+R_{chem}) for LSCF64 is less than that for LSCF82, the activation energy value of the former is 112 - 135 kJ/mol (~1.3 eV) and that of the latter is 156 - 164 kJ/mol (~1.65 eV). The activation energy value is in reasonable agreement with those reported in the literature ~1.4 eV [25,29], or ~1.5 eV [26,27].

Figure 9 presents the temperature dependence of time constant t_{chem} for non-charge-transfer processes of the ORR. A small ORR time constant means a high catalytic activity toward the ORR. Hence, LSCF64 is generally a better electrocatalyst, compared with LSCF82. The activation energy value for LSCF64 is 97 - 121 kJ/mol. In contrast, the activation energy value for LSCF82 is 131 - 173 kJ/mol. The minimum t_{chem} and R_{chem} values of LSCF6482 ensure its superior ORR activity among the LSCF compositions.

A single chamber design based on the above cathode configuration was prepared and tested using methane as fuel and air as oxidant. The single chamber SOFC adopted LSCF6482 as the cathode composition, GDC plus Ni as the anode, and SDC protected LDMW82 as the electrolyte. The methane:oxygen ratio was fixed at 1:1. The current-voltage curves of 550, 600, 650 °C are plotted in Figure 10, along with the specific power curve. The best performance was measured at 600 °C, with the maximum power density reached 126 mW cm^{-2}. We continue working on this SOFC based on the LAMOX electrolyte, optimistic about its improvement in the future.

ACKNOWLEDGEMENT

The authors acknowledge the financial support of the National Science Council of Taiwan through NSC 96-2221-E011-164-MY3.

Table I Values of characteristic resistance R_{chem}, (R_I+R_{chem}) and time constant t_{chem} for the half cells with a SDC interlayer, measured at 700 °C. The electrolyte is LDMW82.

LSCF	R_{chem} (Ω cm^2)	t_{chem} (s)	R_I+R_{chem} (Ωcm^2)
6428	2.89	0.0216	4.54
6446	0.57	0.0108	1.25
6464	0.20	4.34×10^{-3}	0.80
6482	0.15	4.28×10^{-3}	0.77
8228	4.69	0.0587	14.34
8246	4.10	0.0508	11.44
8264	4.27	0.0495	10.81
8282	4.21	0.0214	8.75

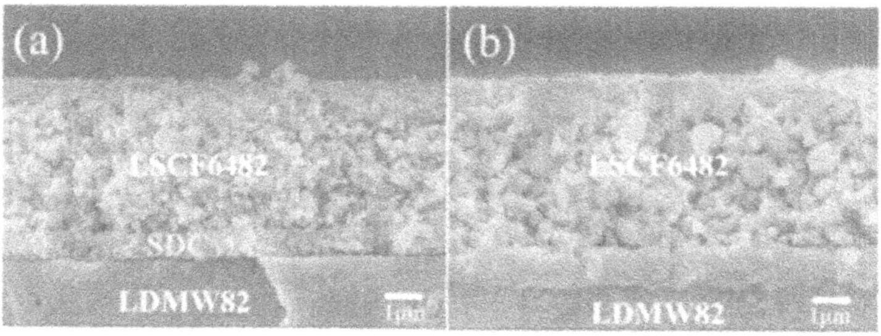

Figure 1 SEM micrographs at the interface of LSCF6482/LDMW82 annealed at 800 °C for 6 h; (a) with a 0.6 μm SDC interlayer and (b) without the SDC interlayer. Note the influences of solid state reaction in (b).

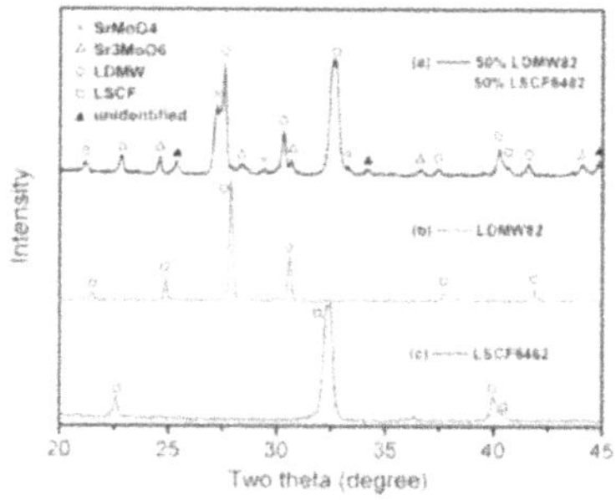

Figure 2 XRD patterns of (a) a powder mixture consisting of 50 wt% LDMW82 and 50 wt%

LSCF6482, (b) LDMW82, (c) LSCF6482.

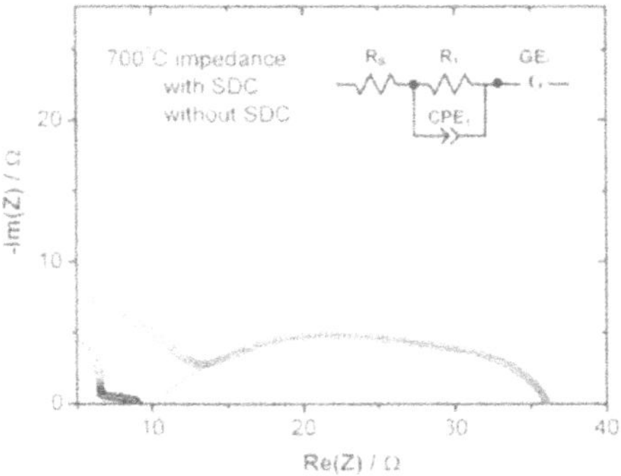

Figure 3 Comparison of two impedance spectra of the as-prepared cells LSCF6482/SDC/LDMW82 and LDMW6482/LDMW82, measured at 700 °C. Note the impedance arc with SDC is much smaller than that without SDC. The equivalent circuit is used for data fitting throughout this work.

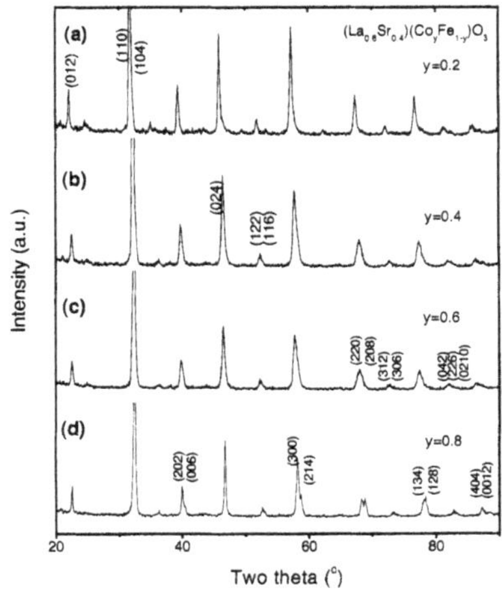

Figure 4 XRD patterns of 1000 °C calcined LSCF powder, containing 40% Sr and (a) 20%, (b) 40%, (c) 60%, (d) 80% Co content.

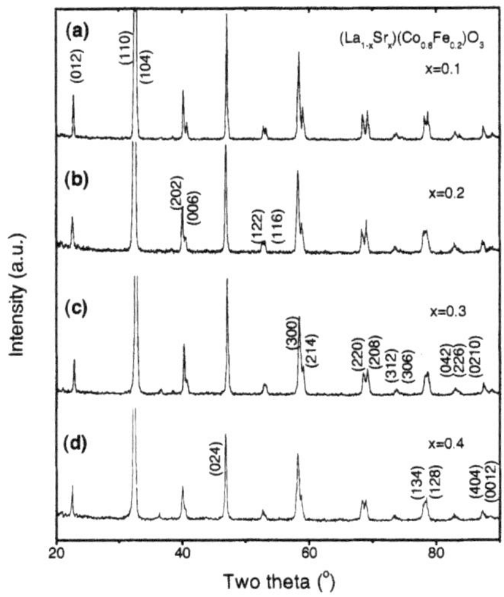

Figure 5 XRD patterns of 1000 °C calcined LSCF powder, containing 80% Co and (a) 10%, (b) 20%, (c) 30%, (d) 40% Sr.

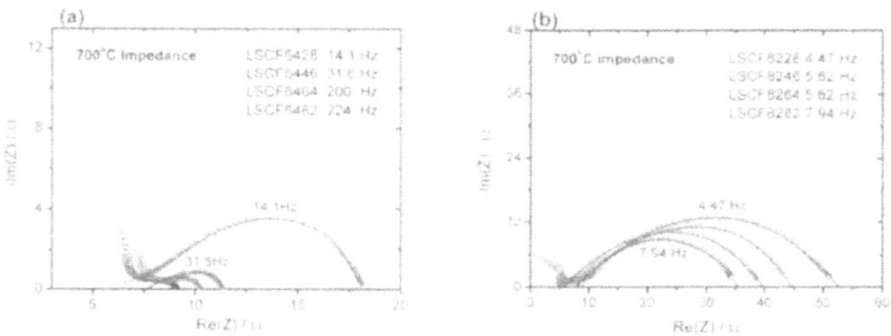

Figure 6 Nyquist plots of 700 °C impedance for the symmetric cells of (a) LSCF6428, 6446, 6464, 6482; (b) LSCF8228, 8246, 8264, 8282 on SDC/LDMW82.

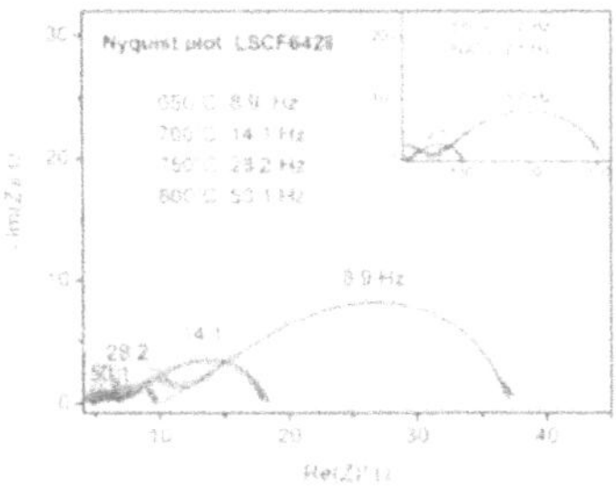

Figure 7 Variation in the impedance arc of LSCF6428/SDC/LDMW82 symmetric cell with temperature.

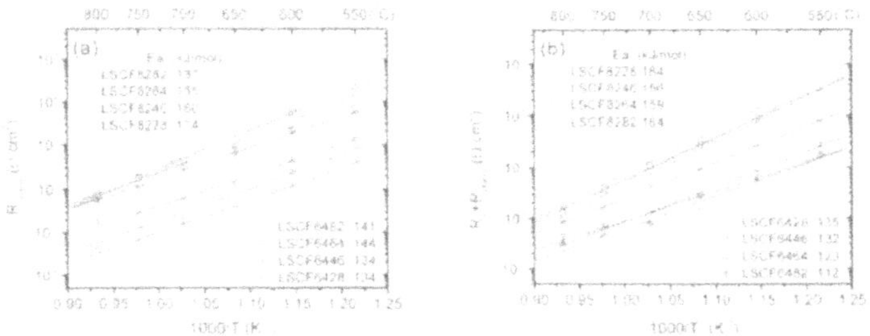

Figure 8 Arrhenius plots of (a) R_{chem} and (b) (R_1+R_{chem}) for two LSCF series, one contains 40% Sr (LSCF64) and the other contains 20% Sr (LSCF82). Also marked are the activation energy values.

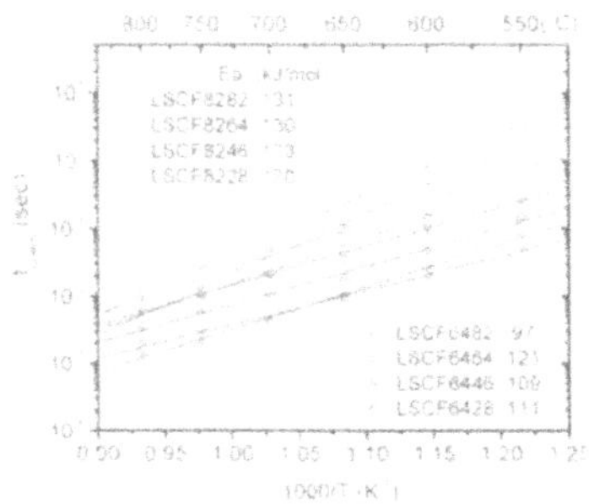

Figure 10 Arrhenius plots of time constant t_{chem} for two LSCF series, one contains 40% Sr (LSCF64) and the other contains 20% Sr (LSCF82). Also marked are the activation energy values for time constant.

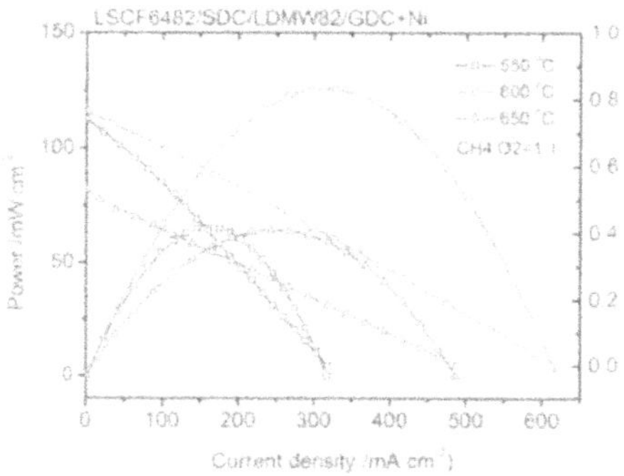

Figure 10 *I-V* and specific power density of a single chamber SOFC operated at 550, 600, 650 °C. The cathode composition and the interlayer are based on the half-cell study.

CONCLUSION

A porous SDC layer derived from the citrate gel method has been implemented to obstruct the cation diffusion at the interface of LSCF/LDMW. The half cell study of LSCF cathode built on LDMW82 protected by a thin SDC layer shows that its performance is comparable to those reported in the literature. The impedance results of two series of LSCF solid solutions containing 20 and 40% Sr are recorded and compared. The LSCF cathodes with 40% Sr are superior to those with 20% Sr, especially the cathodes of LSCF6482, 6464, 6446. Among the compositions being examined, LSCF6482 demonstrates the minimum polarization resistance. A constructed single chamber cell based on this cathode also confirms its validity.

REFERENCES

[1]C.S. Tedman, Jr., H.S. Spacil, S.P. Mitoff, *J. Electrochem. Soc.* **116**, 1170 (1969).

[2]H. Yokokawa, *Annu. Rev. Mater. Res.* **33**, 581 (2003).

[3]G.Ch. Kostogloudis, Ch. Ftikos, *J. Euro. Ceram. Soc.* **18**, 1707 (1998).

[4]F. Goutenoire, O. Isnard, R. Retoux, P. Lacorre, *Chem. Mater.* **12**, 2575 (2000).

[5]P. Lacorre, F. Goutenoire, O. Bohnke, R. Retoux, Y. Laligant, *Nature (London)* **404**, 856 (2000).

[6]F. Goutenoire, O. Isnard, E. Suard, O. Bohnke, Y. Laligant, R. Retoux, P. Lacorre, *J. Mater. Chem.* **11**, 119 (2001).

[7]D. Marrero-Lopez, J. Pena-Martinez, J.C. Ruiz-Morales, D. Perez-Coll, M.C. Martin-Sedeno, P. Nunez, *Solid State Ionics* **178**, 1366 (2007).

[8]G. Corbel, P. Lacorre, *J. Solid State Chem.* **179**, 1339 (2006).

[9]T.Y. Jin, D.S. Tsai, W.H. Chung, F.J. Liu, *J. Am. Ceram. Soc.* **91**, 2217 (2008).

[10]C.W. Chen, D.S. Tsai, T.Y. Jin, W.H. Chung, C.C. Chou, *Solid State Ionics* **179**, 330 (2008).
[11]G. Corbel, S. Mestiri, P. Lacorre, *Solid State Sci.* **7**, 1216 (2005).
[12]N. Jordan, W. Assenmacher, S. Uhlenbruck, V.A.C. Haanappel, H.P. Buchkremer, D. Stover, W. Mader, *Solid State Ionics* **179**, 919 (2008).
[13]H.G. Jung, Y.K. Sun, H.Y. Jung, J.S. Park, H.R. Kim, G.H. Kim, H.W. Lee, J.H. Lee, *Solid State Ionics* **179**, 1535 (2008).
[14]S.P. Simner, J.F. Bonnett, N.L. Canfield, K.D. Meinhardt, J.P. Shelton, V.L. Sprenkle, J.W. Stevenson, *J. Power Sources* **113**, 1 (2003).
[15]A. Mai, V.A.C. Haanappel, F. Tietz, D. Stover, *Solid State Ionics* **177**, 2103 (2006).
[16]JCPDS card no. 85-0586 ($SrMoO_4$), JCPDS card no. 28-1235 (Sr_3MoO_6).
[17]S.B. Adler, *Solid State Ionics* **111**, 125 (1998).
[18]S.B. Adler, *Solid State Ionics* **135**, 603 (2000).
[19]S.B. Adler, J.A. Lane, B.C.H. Steele, *J. Electrochem. Soc.* **143**, 3554 (1996).
[20]S.B. Adler, *Chem. Rev.* **104**, 4791 (2004).
[21]E.P. Murray, M.J. Sever, S.A. Barnett, *Solid State Ionics* **148**, 27 (2002).
[22]L.W. Tai, M.M. Nasrallah, H.U. Anderson, D.M. Sparlin, S.R. Sehlin, *Solid State Ionics* **76**, 259 (1995).
[23]L.W. Tai, M.M. Nasrallah, H.U. Anderson, D.M. Sparlin, S.R. Sehlin, *Solid State Ionics* **76**, 273 (1995).
[24]M.H.R. Lankhorst, H.J.M. Bouwmeester, *J. Electrochem. Soc.* **144**, 1268 (1997).
[25]J. Liu, A.C. Co, S. Paulson, V. I. Birss, *Solid State Ionics* **177**, 377 (2006).
[26]F. Deganello, V. Esposito, M. Miyayama, E. Traversa, *J. Electrochem. Soc.* **154**, A89 (2007).
[27]S. Liu, X. Qian, J. Xiao, *J. Sol-Gel Sci. Technol.* **44**, 187 (2007).
[28]S. Bebelis, N. Kotsionopoulos, A. Mai, F. Tietz, *J. Appl. Electrochem.* **37**, 15 (2007).
[29]M. Hori, K. Nagasaka, M. Miyayama, G. Trunfio, E. Traversa, *Key Eng. Mater.* **301**, 155 (2006).

Ceramics for Electric Energy Generation, Storage and Distribution

ENVIRONMENTAL IMPACT EVALUATION OF CERAMICS USING EXERGY ANALYSIS

Hideki Kita, Hideki Hyuga, and Naoki Kondo
National Institute of Advanced Industrial Science and Technology (AIST)
2266-98 Shimo-Shidami, Moriyama-ku, Nagoya 463-8560, Japan

ABSTRACT

Structural ceramics parts have high potential for the contribution to sustainable manufacturing.For the widespread use, it is important to consider the value of reducing the impact of a product in terms of natural resource consumption and environmental impacts throughout its life cycle. In order to achieve this, we need a method for quantitative evaluation of natural resource consumption similar to measurements of economic cost. Based on these considerations, we carried out an exergy analysis for ceramics in order to clarify the value of environmental impact reduction throughout their life cycles. In this paper, we estimated the environmental impact in the production and usage stages when ceramic heat-tubes were used, compared to conventional steel heat-tubes.

INTRODUCTION

Although structural ceramics have excellent properties compared to other metals, they have not been used widely because of the high initial cost and uncertainties in their fracture behavior due to its brittleness. For their widespread use, the continuous R&D for the cost reduction and improvement of the properties are desired to be conducted.

On the contrary, today, environment and economy must both be sustained, it is also important to consider the value of reducing the impact of a product in terms of natural resource consumption and environmental problems throughout its life cycle. In order to achieve this, we need a method for quantitative evaluation method for natural resource consumption that is similar to measurements of economic cost. In this regard, we believe that exergy analysis could be an effective approach.

Examples of exergy analysis that have been carried out to date for artificial products include research into exergy consumption in the usage stage of electrical lighting, in the production and consumption of reinforced concrete, and in the life cycle of aluminum and steel cans; among natural products, analysis of exergy consumption in plants during photosynthesis has been carried out [1)-6)]. And studies using exergy analysis on ceramic related issues have not been found.

Based on these considerations, we carried out exergy analysis for ceramics in order to clarify the value of environmental impact reduction throughout their life cycles and to develop a design policy for a highly efficient ceramics production process[7)8)].

In this report, we calculated the exergies for natural resources and energies in the production and usage stages when ceramic heater protection tubes were used, compared to conventional steel heater protection tubes, in order to estimate the reduction in an environmental impact.

THE OBJECT FOR ANALYSIS

Aluminum casting line operation and role of heater tube

Aluminum has excellent heat conductivity and is lightweight, and therefore is used widely in engine parts. Also, aluminum is highly recyclable, so disposed engines are collected as scraps and recycled as engine after undergoing some processes. Figure 1 shows the recycling system for aluminum casting line operation.

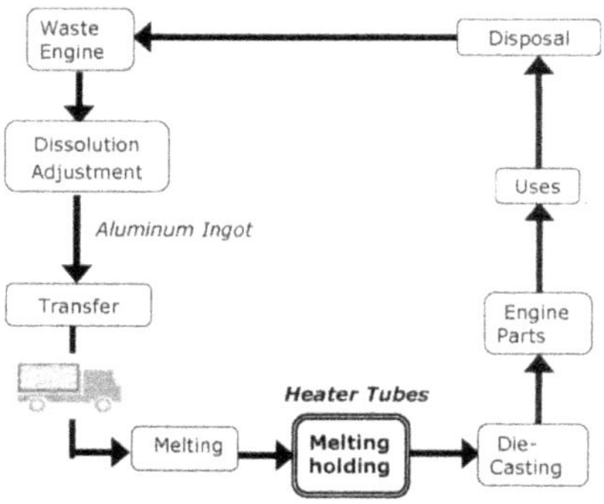

Fig. 1 Recycling System for Aluminum

First, recovered waste engine (scrap) are melted in centralized furnace. They are made into solid ingots, delivered within the plant, melted again in centralized furnace, and then transferred to holding furnace. The molten metal is adjusted for temperature and content, distributed to die cast machine, and formed into products. In this cycling system, there are many factors that decrease efficiency including heat loss, oxidation of molten aluminum, and inclusion of impurities. Input of energies and things from outside is unavoidable to maintain certain quality level and production volume, and reducing these inputs is expected to increase the efficiency of the cycling system.

In the melting and casting production of aluminum, silicon nitride ceramics are now being applied as many components exposed into the molten aluminum due to their superior corrosion and thermal-shock resistances. As one of measure, use of ceramics in production member has been attempted. Heater tube (Figure 2) used in holding furnace is one example. It is a protective tube that envelops the heating wires, and is used to maintain constant temperature of molten aluminum. Heat efficiency increases by using highly conservative silicon nitride, which allows horizontal dip structure where the tube is fixed horizontally in the bottom of the furnace. However, ceramic tube is much more expensive than iron tube. We conducted exergy analysis for manufacture-use-disposal in cases where the heater tube (weight 19 kg) was made with silicon nitride and when it was made with iron.

We made exergy analyses on production of heater protection tubes. Data for its production (quantities of raw material, fuel, waste, waste gas, water, etc.), which were necessary for analyzing exergy, were obtained from companies [9]. Some data were unavailable, and they were supplemented with the author's knowledge of ceramic processing.

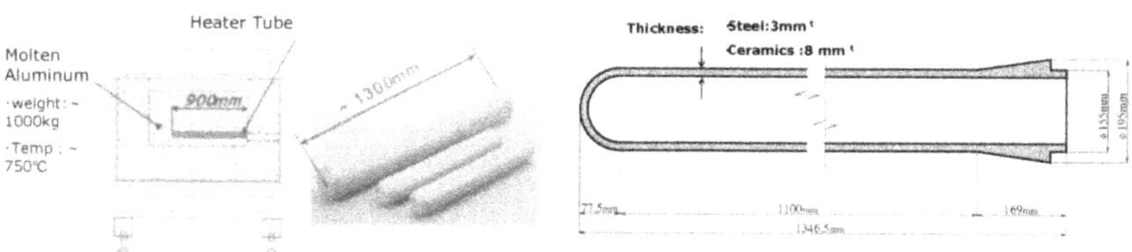

Fig.2 Structure of Melting Furnace and the Dimension of Heater Protection Tube[8]

ANALYSIS METHODS

Exergy Calculations

①Chemical Exergy for Metal and Inorganic Compounds [10]

Referential species of inorganic compounds are denoted Xx, Aa, Bb, and so on. They are formed according to the following reaction formula.

$$xX + aA + bB + \cdots \rightarrow X_x A_a B_b \cdots \quad \cdots\cdots(1)$$

Using $\triangle G^0$ for the Gibbs free energy change, the chemical exergy of the inorganic compound can be calculated using the following formula.

$$E_x{}^0 = \frac{1}{x}\left[-\Delta G^0 - aE_x{}^0(A) - bE_x{}^0(B) - \cdots\right] \quad \cdots\cdots(2)$$

As a reference, the exergy value of a substance is defined as zero if it does not react in a temperature environment of 25 °C (298.15 K). References for some artificial materials are published in *Japanese Industrial Standards* (JIS) [11] , but for those not mentioned in JIS, the reference is the smallest free energy value[12].

②Organic material

Although equations of Rant[8] and Szarut[9] are known for calculation of chemical exergy for organic material, we used the following equation[10] derived by Nobusawa *et al* who modified the equations for practical use.

$$Ex = m \cdot H_l \cdot \left(1.0064 + 0.1519\frac{\phi_H}{\phi_C} + 0.0616\frac{\phi_O}{\phi_C} + 0.0429\frac{\phi_N}{\phi_C}\right) \cdots\cdots(3)$$

m and H_l are dry mass (kg) and lower heating value (J/kg) of organic compounds respectively. φ_C, φ_H, φ_O, and φ_N are weight fractions of carbon, hydrogen, oxygen, and nitrogen in the organic compound.

③Electric power and gas fuel

Electric power is energy that does not contain entropy, so it was used as value for exergy. On the other hand, exergy of fuel gas was calculated using the following equation[10].

$$e_c^o = \sum x_i e_{ci}^o + R'T_o \sum x_i \text{kcal/m}_n^3 \quad \cdots\cdots(4)$$

e_c is exergy and superscript *0* indicates standard temperature (25°C), and subscript *i* is for ingredient *i*. x_i is volume fraction of ingredient *i*.

Production Subsystems[13]

For the calculation of exergy input, consumption, and waste, the production system was assumed to be an assembly of subsystems, and the quantification was performed for the input/output exergy of each production subsystem. An example is shown in Figure 3. For each subsystem, raw materials, fossil fuels (including electric power and steam), etc. containing exergy were inserted (input), and waste, waste heat, etc. as well as a product or an intermediate product containing exergy were generated (output). The obtained intermediate product was used as the starting material for the next subsystem.The final product was obtained by going through these subsystems, and calculating the exergy values that enter or exit the respective subsystems.

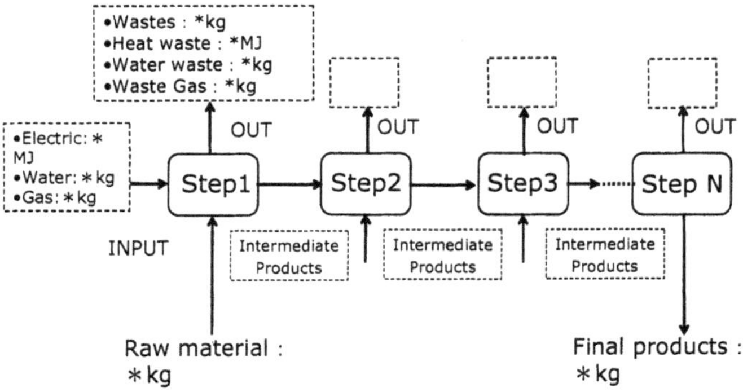

Fig.3 Assembly of Subsystems for the Process

RESULTS AND DISCUSSION[13)]

Calculation of chemical exergy

In conducting the analysis, it is necessary to calculate the exergies of all materials involved in the manufacture. The process of calculation of exergy is shown using silicon nitride (Si_3N_4), the main material. The reference material of silicon nitride is silica and air.

$$Si+(2/3)N_2=(1/3)Si_3N_4 \quad \cdots\cdots(5)$$
$$Ex(Si_3N_4)=3(\Delta G^0)+3Ex(Si)+2Ex(N_2) \quad \cdots\cdots(6)$$
$$Ex(N_2)=RT_0 ln(101.3/76.57) \quad \cdots\cdots(7)$$
$$Ex(O_2)=RT_0 ln(101.3/20.61) \quad \cdots\cdots(8)$$
$$Ex(Si)= (-\Delta G^0) +Ex(SiO_2) -Ex(O_2) \quad \cdots\cdots(9)$$

Ex(Xi) shows the exergy for material Xi. The section in parenthesis in equation (8) is ratio of total pressure of air and partial pressure of nitrogen. From the above equations, exergy of silicon nitride was calculated to be 1877 kJ/mol. Using similar method, exergies values for major fuels were calculated (Table 1).

Exergy analysis at each process

①Production

The exergy balance for the production of a steel protection tube is shown in figure 4. Based on the incoming and outgoing materials and energies for the respective processes, the exergy input for one tube was calculated to be 621.0 MJ. Of this exergy input, 126.0 MJ was fixed in the product; thus, the remainder of 495.0 MJ is considered to be disposed of as waste or heat. Then, the process efficiency of steel was approximately 20%.

Table 1 Chemical Exergy for the Materials and Fuels

Materials or Fuel	Value of Exergy
Y_2O_3	47.0 kJ/mol
Al_2O_3	0 kJ/mol
Si_3N_4	1876.8 kJ/mol
N_2	0.7 kJ/mol
CO_2	20.1 kJ/mol
O_2	3.9 kJ/mol
PVA	49.4MJ/kg
LPG	48.3MJ/kg

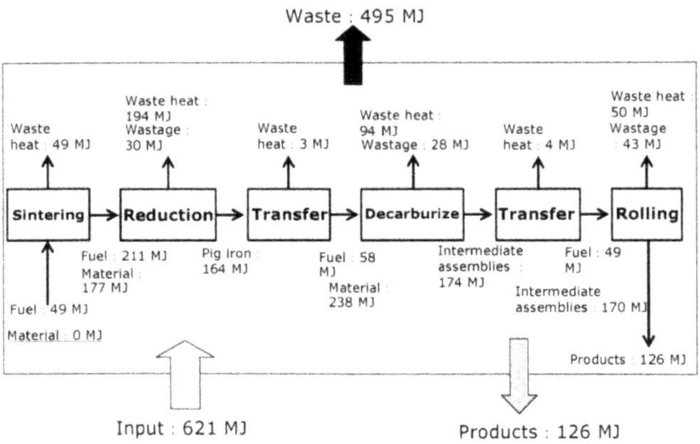

Fig.4 Exergy Balance for the Production of Steel Protection

Figure 5 shows the exergy balance for the production of a ceramic heater protection tube[7)8)]. The process involves mixing, granulation, CIPing and green machining, dewaxing, and sintering. As the input energy, LPG was used for drying in granulation, and electric power was used in dewaxing and sintering. Most of the material removed by green machining is supposed to be reused; thus, the raw material powder is not wasted. When the same weight of silicon nitride as the above was produced, the total input exergy of 4175 MJ, which is the sum for raw materials, electric power, and gas (312 MJ, 1617 MJ, and 2247 MJ, respectively), was necessary. The exergy fixed in the product was 229 MJ, and the remaining 3946 MJ was discharged into the environment as waste or heat. The process efficiency of ceramics was calculated by Ex(products) / Ex(input) = 229 MJ / 4175 MJ, then it was approximately 5.5%.

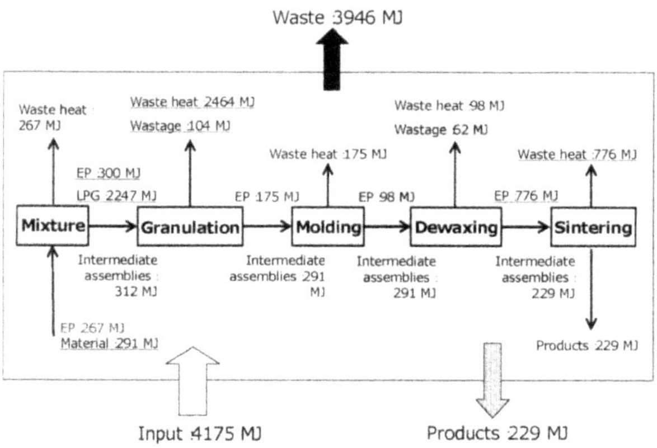

Fig. 5 Exergy Balance for the Production of a Ceramic Heater Protection Tube

Use

① Wear and material disposal

When steel heater tube is used in molten aluminum, it is corroded by aluminum and worn down by passage of time *t*. It was assumed that wastage progressed according to the following equation.

$D=Do\times(2\text{-}exp(kt))$ ····· (11)

Here, D is thickness of heater tube (mm), D_0 is initial thickness (mm), k is apparent reaction speed coefficient, and D_i is thickness at time of replacement (mm). Assuming D_0 = 3 mm (from data) and D_i = 0.5 mm, under condition of replacement every half year, the reaction constant k was calculated to be 0.07.

Consumption exergy is expressed by following equation.

$E=Eo\times exp(kt)$ ····· (12)

Exergy of steel is 6.6 MJ/kg (= 368 KJ/mol) and total weight of the product is 19 kg, and when it is disposed when damage reaches Di, consumption is 126 MJ/tube. While steel heater tube is exchanged once every half year, silicon nitride is stable and does not react, and is exchanged and disposed along with furnace that has lifespan of 7 years. Figure 9 shows the change with time of exergy consumption in 7 years. Exergy consumed by disposal during this time is as shown in the following equation.

- Steel: 126 (MJ/tube) × 14 (tubes) = 1764 MJ
- Silicon nitride: 229 (MJ/tube) × 1 (tube) = 229 MJ

When steel heater tube is used, damage and disposal are repeated and exergy consumption increases in step-like form. In contrast, there is hardly any consumption in 7 years using ceramics, and exergy value (229 MJ) is released at the end of furnace lifespan. Also, using ceramics, there is less chance of inclusion of impurities compared to steel, so clean molten metal can be obtained, and this is another advantage of ceramics.

② Running

(a) Melting and holding furnace

In vertical dip type using steel heater tube, 9.4 kW is required during run and 4.0 kW at rest, while in horizontal dip type using silicon nitride, electricity consumptions at run and rest are 6.8kW and 3.8 kW respectively, due to improved heat efficiency. While it will be running 60% (40% rest) per day and is in operation 360 days a year, the total electricity consumed in 7 years, or exergy input, will be as follows.

- Steel: (9.4x0.6x24+4.0x0.4x24) × 360 × 7 × 3.6/1000 = 1576 GJ
- Silicon nitride: (6.8x0.6x24+3.8x0.4x24) × 360 × 7 × 3.6/1000 = 1219 GJ

(b) Die cast machine

Assuming that electricity consumption of die cast machine is 20 kW, running 60% per day for 360 days per year, the total electricity consumption, or exergy input, for 7 years is as follows.

- 20 × 0.6 × 24 × 360 × 7 × 3.6/1000 = 2612 GJ

③Manufacture, use, and disposal

As result of interview with companies, the total manufacture volume of cast product in 7 years was estimated to be about 4300 ton. In this paper, material loss is not considered. Therefore, the amount of molten aluminum is 4300 ton or same as final product, and the exergy was calculated as 126802 GJ in molten condition (temperature 700°C), and 125582 GJ in solid condition.

Figure 10 shows the amount and flow of exergy input and output for manufacture using ceramics and steel heater tube, their use in melting and holding furnace when casting was conducted for 7 years. As mentioned above, when the furnace is run for 7 years, 14 steel tubes are required since they are subject to damage. Therefore, energy input and output during the manufacture process is as follows.

- Input: 621 (MJ/tube) × 14 (/tubes) = 8694 MJ
- Output: 495 (MJ//tube) × 14 (/tubes) = 6930 MJ

On the other hand, only one silicon nitride tube is required during same time, and exergy for input and output will be 4175 MJ and 3946 MJ respectively according to Figure 7. Next, exergy accompanying damage and disposal during use is as follows.

- Steel: 126 (MJ/ tube) × 14 (tubes) = 1764 MJ
- Silicon nitride: 229 (MJ/ tube) × 1 (tubes) = 229 MJ

Looking over the entire process, the exergy inputs for steel and silicon nitride were 130999 GJ and 130637 GJ respectively, while exergy outputs were 5417 GJ and 5055 GJ. Using silicon nitride reduced 362 GJ of input and output exergy compared to steel.

From the above results, it was shown that although one silicon nitride tube required 7 times more exergy in manufacturing process, frequency of replacement decreased due to its high conservative property, which allowed furnace with highly efficient structure that reduced electricity consumption, and therefore, exergy consumption level was smaller compared to steel in total throughout the lifecycle of manufacture, use, and disposal.

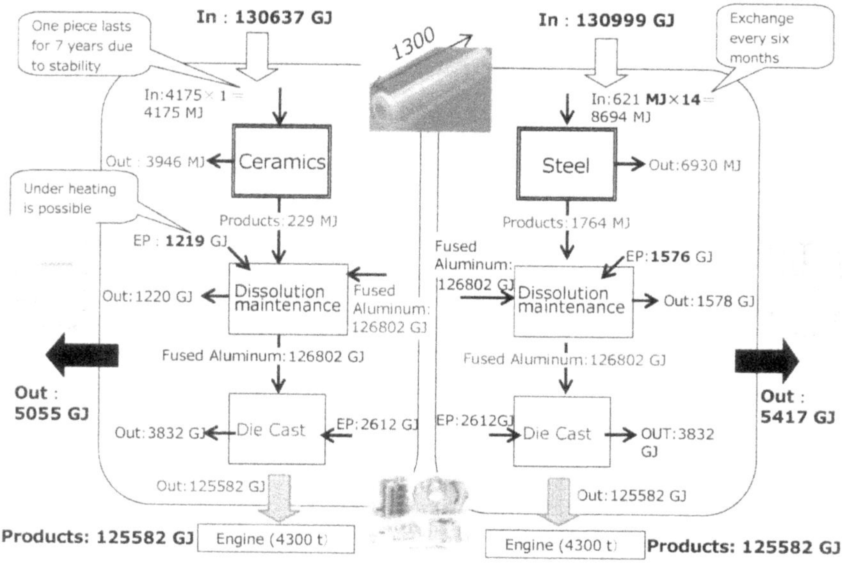

Fig.6 Consumed Exergy through Production of the Heater tubes, their Use and Disposal[13)]

CONCLUSIONS

Exergy analysis was done in case where members were manufactured with ceramics and steel, used as production member for 7 years in casting line for aluminum engine, and then disposed.

①Exergy input per product was 4175 MJ for ceramics and 621 MJ for steel, and significantly higher exergy was consumed for ceramics compared to steel.
Exergy fixed in ceramics was 229 MJ, or 5.5% of exergy input, and most are wasted outside the system.

②As result of high exergy input, ceramics have high conservation. When it is used in molten aluminum using this characteristic and used for 7 years, exergy consumption is reduced by 362 GJ compared to steel.

REFERENCES

[1] H. Asada, and M. Shukuya, "Exergy-Entropy Process of Electric Lighting Systems Using Fluorescent Lamps", *J. Archit. Plann. Environ. Eng.*, 483, 91-100(1996) , 【in Japanese】
[2] M. Saito, I. Takahashi, and M. Shukuya, "Numerical Analysis of Exergy Balance of A Leaf -Research on the Passive Cooling Effect of Trees", *Archit. Plann. Environ. Eng.*, 505, 51-58(1998) , 【in Japanese】
[3] I. Takahashi, and M. Shukuya, "An Exergy Analysis of the Circulation of Matter Obtained from Composing Garbage in an Urban Area", *J. Archit. Plann. Environ. Eng.*, 510, 23-29(1998) , 【in Japanese】
[4] J. Dewulf and H. R. Langenhove, "Quantitative Assessment of Solid Waste Treatment Systems in the Industrial Ecology Perspective by Exergy Analysis", Environ. *Sci. Tech.*, 36, 1130-1135(2002)
[5] T. Morihana, I. Takahashi, and M. Shukuya, "Exergy Consumption of Reinforced Concrete Walls in the Courses of Production and Use", *J. Archit. Plann. Environ. Eng.*, 520, 99-105(1999) , 【in Japanese】
[6] Y. Soeno, T. Akashi, H. Ino, K. Shiratori, K. Nakajima and K. Harada, "Exergy Analysis for the Integrated Evaluation of Environmental Impacts", *J. Japan Inst. Metals*, 66[9], 885-888(2002) , 【in Japanese】
[7] H. Kita, et al. "Exergy Analysis on the Ceramic Manufacturing Process", *J. Ceram. Soc. Japan*, 115[12] 987-992(2007) 【in Japanese】
[8] H. Kita, H. Hyuga, N. Kondo, T. Ohji, " Exergy Consumption Through The Life Cycle of Ceramic Parts", *International Journal of Applied Ceramic Technology*, 5[4]' 373-381 (2008)
[9] Private Communication
[10] K. Karakida, "Ekuserugi no kiso", Ohm-sya (2005) , 【in Japanese】
[11] "General Rules for Energy Evaluation Method by Available Energy", *Nihon Kogyo Kikaku*, Z 9204, (1991) , 【in Japanese】
[12] e.g. MALT (Netsuriki-Gaku Data Base), *Kagaku Gijyutsu-Sya*
[13] H. Kita, et al. "Guideline for rationalization of resource and energy use considering the complete process of manufacturing ", *Synthesiology*, 5[3], 212 - 221 (2008)

THE DYNAMIC CHARACTERISTICS OF ULTRASONIC THERAPEUTIC TRANSDUCERS USING LEAD-FREE $Na_{0.5}K_{0.5}NbO_3$ BASED CERAMICS

Ming - Ru Yang [a], Sheng - Yuan Chu[a, b, *], Cheng - Che Tsai[c], Cheng - Shong Hong[d] and Chih-Lung Lin[a]

[a] Department of Electrical Engineering, National Cheng Kung University

[b] Center for Micro/Nano Science and Technology, National Cheng Kung University

[c] Department of Electronics Engineering and computer science, Tung-Fang Institute of Technology

[d] Department of Electronic Engineering, National Kaohsiung Normal University

ABSTRACT:

In this paper, lead-free $Na_{0.5}K_{0.5}NbO_3$ (NKN) based piezoelectric ceramics are synthesized by the conventional mixed-oxide methods with sintering temperature at 900-1050 °C. Microstructural analyses of un-doped and CuO-doped ceramics have been carried out and compared using X-ray diffraction (XRD) and scanning electron microscope (SEM). The dependence of the sintering temperatures of the CuO-doped NKN ceramics on the piezoelectric and dielectric properties have been investigated. The results show that the CuO-doped NKN specimens sintered at 1020 °C present the high thickness coupling factor, k_t (>0.50), mechanical quality factor Q_m =1100 and the lower temperature coefficient of resonance frequency in thickness mode ($\frac{df_r}{f_r} < 2\%$) within the temperature range 20-100 °C. Then, we successfully construct 3-MHz transducers (LF transducer) utilizing our fabricated piezoceramics and examine their static resonant characteristics. Simultaneously, their dynamic acoustic power is measured by the acoustic force balance under a power circuit driving with the intermittent pulse control as compared to the commercial transducers. We find the effective coupling factor of LF transducer (0.05) is superior to that of commercial transducer (0.45).

1. INTRODUCTION:

In the past, innovations in actuators and sensors have been investigated and developed in piezoelectric ceramics [1,2]. Modified ceramics have good potential due to their ability to facilitate a desirable combination of properties, such as high surface phase velocity, electromechanical coupling coefficient (k^2) and low temperature coefficient of frequency (TCF) [1,2]. Lead oxide (PbO) based ceramics, such as lead titanate (PT) and lead zirconate titanate (PZT), have been widely studied and used as transducers, piezoelectric actuators, surface acoustic wave (SAW) devices and sensors because of their excellent piezoelectric properties [3-5]. However, its high content of toxic element (Pb>60 wt%) has triggered some concerns in Europe and Japan mainly due to the utilization, recycling, and disposal of lead-based ferroelectrics. Elimination of such issues has become a driving force for developing lead-free ferroelectric ceramics with properties almost comparable to their lead-based counterparts.

Among several candidates for lead-free materials, such as Bi compounds [6,7] and alkaline niobate compounds [8,9], Sodium potassium niobate (($(Na_{0.5}K_{0.5})NbO_3$, NKN) ceramic is an attractive material and has been thoroughly investigated as a result of its high electromechanical coupling coefficient (k^2) and high phase transition temperature (T_c~420° C), especially near the morphotropic phase boundary (MPB) [10–12]. However, NKN exhibits low piezoelectric properties (d_{33} of 80 pC/N, k_p of 36% and Q_m of 40~50) and it is difficult to produce densely by the conventional sintering process. The main problem is the volatilization of potassium oxide (K_2O) at 800° C making it difficult to control the stoichiometry [13,14]. Oxygen deficiency has been another problem in the preparation which results from the high-temperature processing and gives rise to the electronic conductivity [14].Hot-pressing technique as well as spark plasma sintering have been used to obtain high densities [15,16], but such processing techniques are not appropriate for industrial applications. On the other hand, compared with soft NKN-based ceramics, the researches on hard NKN-based ceramics are not sufficient. It is also noted that CuO is an excellent sintering aid and has been used in the improvement in the sintering performances of the piezoelectric ceramics [17,18]. In this study, we continued our previous work and 2 mol% CuO-doped NKN lead-free ceramics were prepared by a conventional ceramic technique and the piezoelectric properties for the ceramics were studied. CuO is used as a dopant to improve the densification, decrease the sintering temperature and hence enhances the properties of NKN ceramics. It is also found that 2 mol% CuO dopants are also helpful to improve the mechanical quality and temperature stability of resonance frequency. Simultaneously, we utilize this developed piezoceramics to fabricate the therapeutic transducers embedded in the human body in the field of physi-therapy applications. According to the authors' knowledge, little research groups have studied on these respects.

2. EXPERIMENTAL PROCEDURE

The starting materials of $K_{0.5}Na_{0.5}NbO_3$ (NKN) and $K_{0.5}Na_{0.5}NbO_3$-2CuO (NKN-2Cu) samples processed by a conventional mixed-oxide method were pure reagent Na_2CO_3 (SHOWA, 99.5%), K_2CO_3 (SHOWA, 99.5%), Nb_2O_5 (SHOWA, 99.5%) and CuO (SHOWA, 99.5%) powders. Then, they were weighed according to the desired compositions. The starting materials were transferred to a 100 mm diameter cylindrical plastic jar, partially filled with 10 mm diameter ZrO_2 grinding balls. Sufficient ethanol (99.5%) was added to cover the powders. Ball milling was carried out for 24 h, followed by drying at 130° C, before grinding using an alumina mortar and pestle to break up large agglomerates formed during drying. In order to enhance the uniformity of the composition, the mixture was calcined at 850° C in air for 5 h. The resulting powders were ball milled for 24 h and dried again. These powders, milled with 5 wt.% PVA aqueous solution, were uni-axially pressed into a disk of 12 mm diameter, at pressure of 25 kg/cm^2 and subsequently sintered in air at 900–1080° C for 3 h.

Bulk densities were measured by the Archimedes method using distilled water as medium. The microstructure was observed by field emission scanning electron microscopy (FESEM) with a Hitachi S-4100 microscope. The crystallographic study was confirmed by X-ray diffraction (XRD) using Cu Kα (λ = 0.154 nm) radiation with a Seimens D-5000 diffractometer operated at 40 kV and 40 mA. The dielectric and piezoelectric properties were measured with a HP 4294A precision impedance analyzer. To measure the electrical properties, silver paste was painted on both sides of the samples to form electrodes, and then subsequently fired at 150◦ C for 20 min. After that, samples were poled under DC field of 40 kV/cm at 150◦ C in silicone oil for 30 min. The electromechanical coupling factors k_p, k_t and mechanical quality factor Q_m were determined by the resonance method according to the IEEE Standard 176 using an impedance analyzer (HP 4294A). The acoustic output power of the devices was measured by acoustic force balance UPM-DT-1AV (Electronic ULTRASOUND POWER METER, Ohmic Instruments Co.)

3. EXPERIMENTAL RESULTS AND DISCUSSIONS

The XRD patterns of the NKN and NKN-2Cu ceramics are shown in Fig. 1. A homogeneous NKN phase was well developed without second phase. However, when CuO adding 2 mol%, the $K_4CuNb_8O_{23}$ (KCN) second phase (JCPDS card No.21-1250), indicated by the asterisk, which was formed in the NKN-2Cu ceramics. A small change in their lattice parameters is found, for NKN, a=5.5719 Å, b=3.9126 Å and c=5.6125 Å, and a=5.5702 Å, b=3.9110 Å and c=5.6113 Å for NKN-2Cu. Their decrease is due to the incorporation of Cu^{2+} into the B-site, and formation of oxygen vacancy in NKN and form a normal ferroelectric.

The NKN-2Cu bulk densities and their relative densities as function of the sintering temperature are shown in Fig. 2. As the sintering temperature increases, the NKN-2Cu bulk density increases with the temperature from 940◦ C to 1020◦ C and reaches maximum value of 4.43 g/cm^3 with 98.2% theoretical density (TD = 4.51 g/cm^3) and is higher than the density of pure NKN ceramics shown in Table I. Fig. 3 shows the microstructures of NKN and NKN-2Cu ceramics. The pure NKN ceramics consist of smaller and uniform grains in comparison with CuO-doped samples. With Cu-addition, a homogeneous microstructure with a larger grain size was formed. Furthermore, a small amount of CuO doping can assist the densification effectively and improve the sintering performance of the NKN ceramics.

The piezoelectric planar and thickness coupling factor, k_p and k_t, were calculated by the following equation [19]:

$$\frac{1}{k^2} = a \times \frac{f_r}{(f_a - f_r)} + b \tag{1}$$

where fr is the resonance frequency, fa is the anti-resonance frequency, a = 0.395 and b = 0.574 for planar (k_p) mode, a = 0.405 and b = 0.810 for thickness (k_t) mode. Longitudinal coupling, k_{33}, was estimated from the piezoelectric planar and thickness coupling factor [3]:

$$k_{33}^2 = k_p^2 + k_t^2 - k_p^2 k_t^2 \tag{2}$$

Dependence of k_p, k_t, k_{33} and Q_m for the NKN-2Cu ceramics on the sintering temperature is shown in Fig. 4. Both k_P and k_t slightly increase with sintering temperature and then decrease, giving a maximum value of 37% and 52%, respectively, at 1020° C. k_{33} increase evidently with sintering temperature from 940° C to 1020° C, reaching 61%, and decrease over 1020° C. The promotion may be attributed to the increased density, lowering the leakage current and enhancing the poling process [19,20]. Moreover, as shown in Table I, the k_t value of the well-sintered NKN-2Cu ceramics is much higher than those of NKN ceramics published and this property is good for transducers.

The values of Q_m is also shown in Fig. 4. The value of Q_m is determined by means of the relation [19]:

$$\frac{1}{Q_m} = 2\pi f_r RC\left(\frac{f_a^2 - f_r^2}{f_a^2}\right) \qquad (3)$$

where R is the resonance impedance and C is the capacitance at 1 kHz. The experimental results show that the Q_m value of NKN-2Cu ceramics is 470 for sintering temperature at 940° C and increases to 1078 for sintering temperature at 1020° C. The resonance impedance of NKN-2Cu were 58 Ω sintered at 940° C and 15 Ω sintered at 1020° C. According to Eq. (3), the Q_m value is inversely proportional to the resonance impedance. Therefore, as the variation of resonance impedance from 58 to15 Ω, the Q_m increased from 470 to 1078. From Table I, the improvement of Q_m value from 124 for pure NKN ceramics to 1078 for NKN-2Cu ceramics may be due to the hardening effect of the Cu ions. Piezoelectric ceramics with "hard" characteristic exhibit high Q_m, low tanδ, and high elastic hardness. In this sense, Cu-doped NKN has the hard characteristic. Therefore, Cu-substitution provides "hard" piezoelectric characteristics for NKN. In the case of PZT, the addition of acceptor ions, such as Mg^{2+}, Sc^{3+}, and Fe^{3+}, increases the density of oxygen vacancy. This leads to enhance the value Q_m and Cu^{2+} is considered to be a substitute for Nb^{5+} and also to behave as an acceptor.

In piezoceramics-based resonant devices such as transducers, resonators and oscillators, the TCF parameter is so important that it is usually used as a key index to evaluate the qualities of these devices. To obtain TCF, samples were placed in temperature-controlled furnace to measure from 20° C (T_1) to 100° C (T_2). The TCF values were calculated by following [21]:

$$TCF = \frac{f_{rT_2} - f_{rT_1}}{f_{rT_1} \times (T_2 - T_1)} \qquad (4)$$

Fig. 5 shows the temperature dependence on the change rates of resonant frequency of pure NKN and NKN-2Cu ceramics. As temperature increase from 20° C to 100° C, the change rates of resonant frequency of NKN-2Cu ceramics is lower than that of NKN ceramics. When the temperature rises to 100° C, the change rates of resonant frequency are calculated as -2.65% for NKN-2Cu ceramics and -6.1% for pure NKN ceramics, and the TCF values of NKN-2Cu and NKN ceramics are -331 ppm/° C and -763 ppm/° C, respectively.

Then, the NKN-2Cu disks are assembled into 3 MHz ultrasonic therapeutic transducers according to the design in Fig. 6. The diameter of NKN-2Cu disk is about 10 mm with thickness of 0.94 mm. Epoxy is used to glue the disks to the aluminum layer and the architecture is shown in Fig. 6(b). As shown in Fig. 7, the setup of ultrasonic therapeutic transducer included self-exciting oscillation circuit, DC power supply, single chip computer and therapeutic transducer. To avoid overheating the therapeutic transducer, the therapeutic transducer is driven by self-exciting oscillation circuit producing a 3 MHz sinusoidal and modulated pulse train with a frequency of 20 Hz such that the therapeutic transducer is operated in on-off condition under self-tuning circuit driving. The plots of impedance and phase for both lead-free transducers (LF transducers) and commercial lead-based transducers are shown in Fig. 8 and the comparison properties of these two types of the transducers also are listed in Table II. The results show that the effective coupling factor of LF transducer (0.05) is superior to that of commercial transducer (0.045), but the acoustic power of LF transducer is the one-fifth lower than that of commercial transducers. It is attributed that the resonance resistance of the LF transducer (72 Ω) is much greater than that of the commercial transducer (8.3Ω). In other words, as the ultrasonic wave entering the NLN-Cu based piezoceramics will cause more energy decay of the ultrasonic wave due to the greater damping factor (i.e. the greater resonance resistance) leading to the acoustic power decreasing. Thus, the resonance resistance of the transducers needs to be lowered down and the developed lead-free compositions to meet the transducer requirements will be continued in the future works.

4. CONCLUSION

In this study, CuO was added to NKN ceramics to decrease the sintering temperature and eventually improve the density and the piezoelectric. The $K_4CuNb_8O_{23}$ (KCN) second phase was formed in the NKN-2Cu ceramics that enhanced the sintering by making liquid phases at sintering temperature. When a small amount of CuO was added, the densities of these samples were found a high value of about 4.43 g/cm^3 with 98.2% theoretical density. The NKN-2Cu ceramics sintered at 1020° C show excellent properties of k_p = 37%, k_t = 52%, k_{33} = 61% and Q_m = 1078. The NKN-2Cu ceramics with high k_t value and very low TCF value (-331 ppm/° C) could be one of the promising lead-free ceramics electromechanical transducers. The ultrasonic therapeutic transducer is successfully fabricated using lead-free NKN-2Cu ceramics and could be operated in the thickness mode.

ACKNOWLEDGEMENTS

This research was supported by the National Science Council of Republic of China, under grant NSC-93-2216-E-006-033. And the authors also gratefully acknowledge the supports from Center for Micro/Nano Technology Research, National Cheng Kung University.

REFERENCES

[1] T.Y. Chen, S.Y. Chu, C.K. Cheng, Inter. Ferro. 58 (2003) 1315–1324.
[2] S.Y. Chu, T.Y. Chen, W. Water, J. Cryst. Growth 257 (2003) 280–285.
[3] B. Jaffe, W.R. Cook, H. Jaffe, Piezoelectirc Ceramics, Academic, (1971) 185–214.
[4] S.Y. Chu, T.Y. Chen, IEEE Trans. Ultrason. Ferroelectr. Freq. Control 51 (2004) 663–667.
[5] S.Y. Chu, T.Y. Chen, I.T. Tsai, W. Water, Sens. Actuator A Phys. 113 (2004) 198–203.
[6] T. Takenaka, T. Gotoh, S. Mutoh, T. Sasaki, Jpn. J. Appl. Phys. 34 (1995) 5384–5388.
[7] M. Hirose, T. Suzuki, H. Oka, K. Itakura, Y. Miyauchi, T. Tsukada, Jpn. J. Appl. Phys. 38 (1999) 5561–5563.
[8] L. Egerton, D.M. Dillon, J. Am. Ceram. Soc. 42 (1959) 438–442.
[9] G.H. Haertling, J. Am. Ceram. Soc. 82 (4) (1999) 797–818.
[10] G. Shirane, R. Newnham, R. Pepinsky, Phys. Rev. 96 (1954) 581–588.
[11] V. Lingwal, B.S. Semwal, N.S. Panwar, Bull. Mater. Sci. 26 (2003) 619–625.
[12] H. Birol, D. Damjanovic, N. Setter, J. Eur. Ceram. Soc. 26 (2006) 861–866.
[13] Y. Guo, K. Kakimoto, H. Ohsato, Solid State Commun. 129 (2004) 279–284.
[14] S.Y. Chu, W. Water, Y.D. Juang, J.T. Liaw, S.B. Dai, Ferroelectrics 297 (2003) 11–17.
[15] R.E. Jaeger, L. Egerton, J. Am. Ceram. Soc. 45 (1962) 209–213.
[16] R. Wang, R. Xie, T. Sekiya, Y. Shimojo, Mater. Res. Bull. 39 (2004) 1709–1715.
[17] E. Li, H. Kakemoto, S. Wada, T. Tsurumi, J. Am. Ceram. Soc. 90 (2007) 1787–1791.
[18] C. W. Ahn, H. C. Song, S. Nahm, S. H. Park, K. Uchino, H. G. Lee, H. J. Lee, J. Am. Ceram. Soc. 89 (2006) 921–925.
[19] M. Matsubara, T. Yamaguchi, K. Kikuta, S. Hirano, Jpn. J. Appl. Phys. 44 (2005) 6136–6142.
[20] S.H. Park, C.W. Ahn, S. Nahm, J.S. Song, Jpn. J. Appl. Phys. 43 (2004) L1072–L1074.
[21] S. Gevorgian, P. K. Petrov, Z. Ivanov, E. Wikborg, J. Phys. Lett. 79 (2001) 1861-1863.

Table I The comparison of properties of $(Na_{0.5}K_{0.5})NbO_3$-$LiTaO_3$ ceramics based on the previous reports of various groups.

Ceramic composition	ρ (g/cm³)	ρ_r (%)	k_t (%)	k_p (%)	Q_m	Reference
NKN(hot pressed)	4.46	98.9	-	45	240	[12,13]
NKN	4.3	95.3	45	39	130	[12]
NKN	4.34	96.2	43	38	124	Our sample
NKN-2Cu	4.43	98.2	52	37	1078	Our sample

Table II Properties of the therapeutic transducers fabricated using our developed NKN-Cu based ceramics as compared to commercial therapeutic transducer made from HES CO. Ltd.

	Resonance frequency (MHz)	Resonance resistance (Ω)	Effective coupling factor (k^2)	Mechanical quality factor (Q_m)	Dynamic electrical power (W)	Dynamic acoustic power (W)
[a]Commercial transducer	3.085	8.32	0.045	25	2.2	0.7
[b]LF transducer	3.030	71.33	0.050	23	2.2	0.15

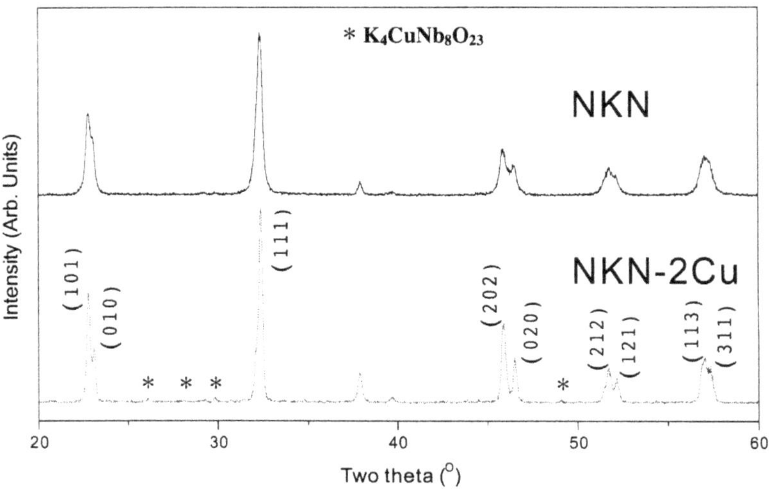

Figure 1 The XRD patterns of the pure NKN and NKN-2Cu ceramics.

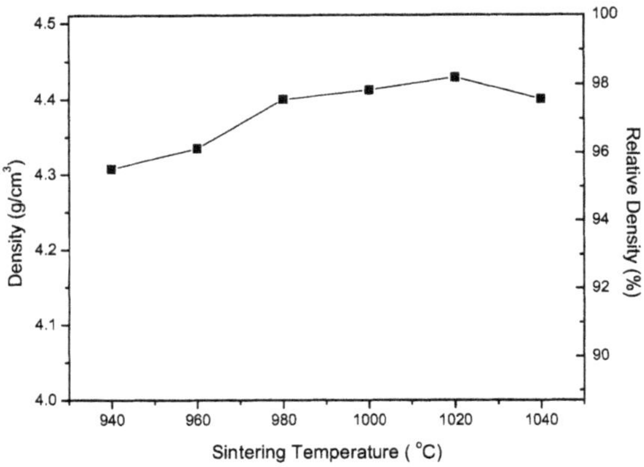

Figure 2 The NKN-2Cu bulk densities and their relative densities as function of the sintering temperature.

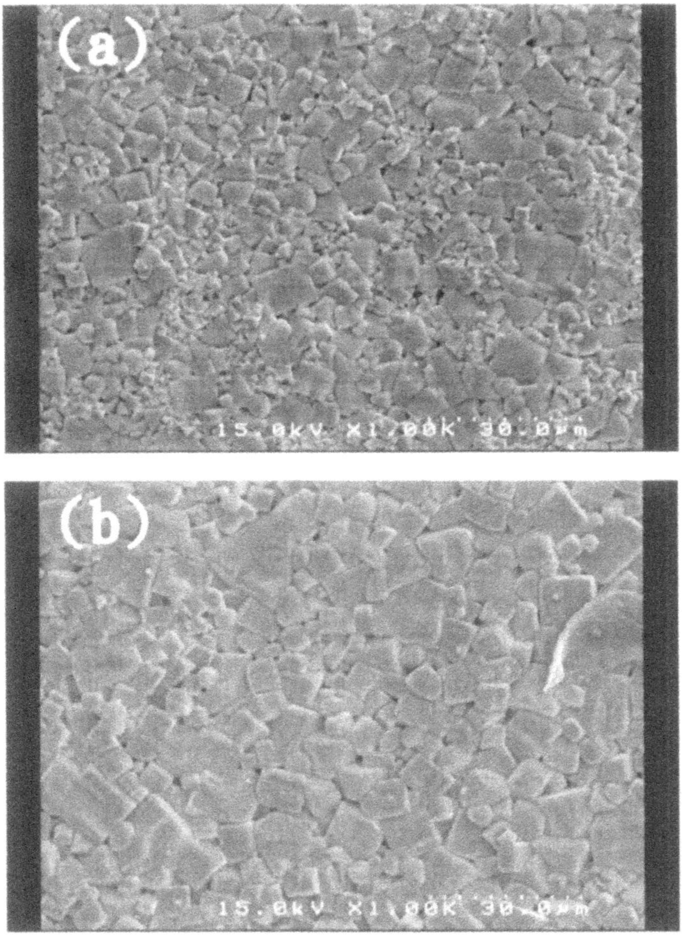

Figure 3 The SEM images of (a)NKN ceramics sintered at 1080◦ C and (b)NKN-2Cu ceramics sintered at 1020◦ C.

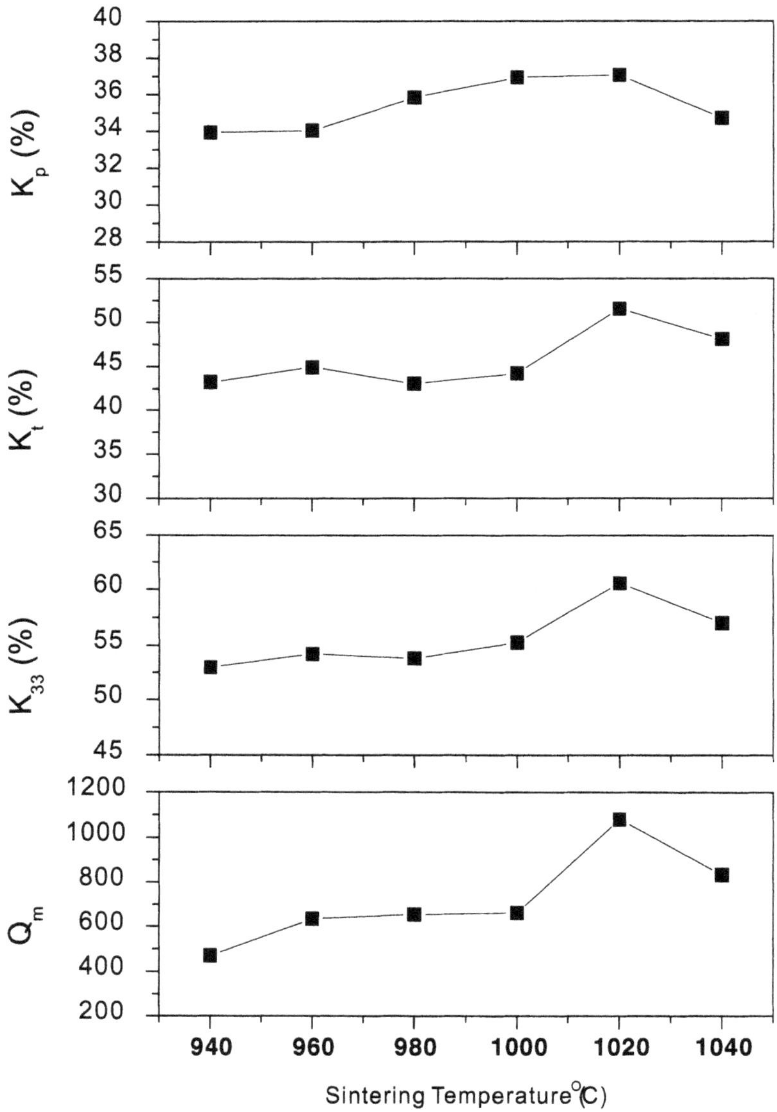

Figure 4 Dependence of k_p, k_t, k_{33} and Q_m values on the NKN-2Cu ceramics sintering temperature.

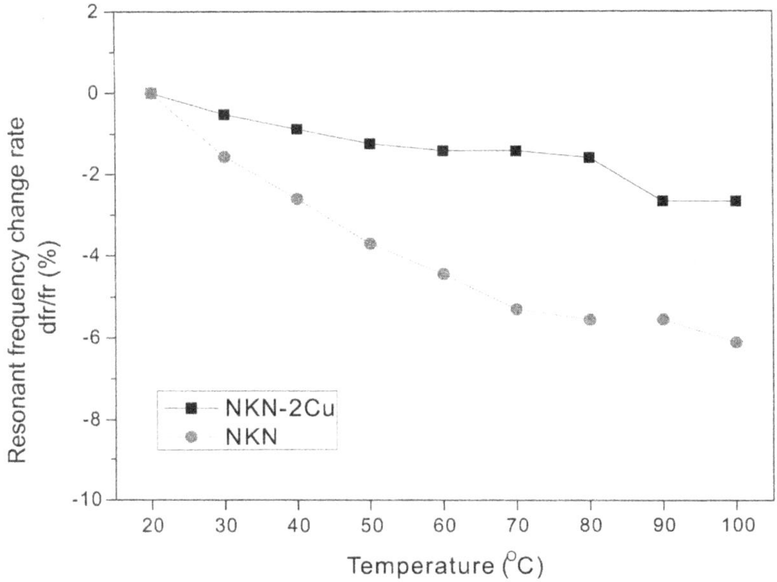

Figure 5 Change rates of resonant frequency of NKN and NKN-2Cu ceramics versus temperature.

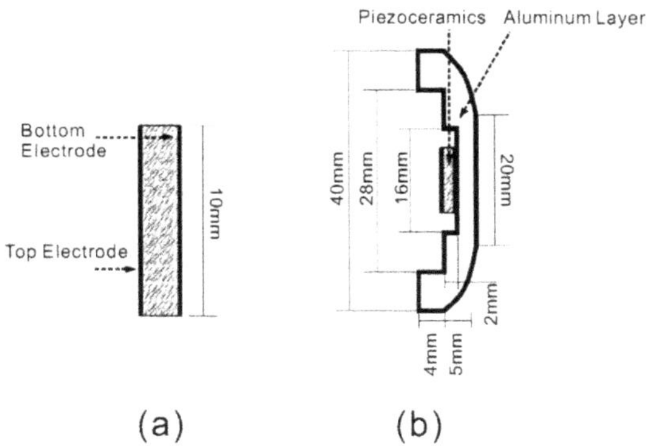

Figure 6 (a) NKN-2Cu disk (b) Architecture of the implemented ultrasonic therapeutic transducer.

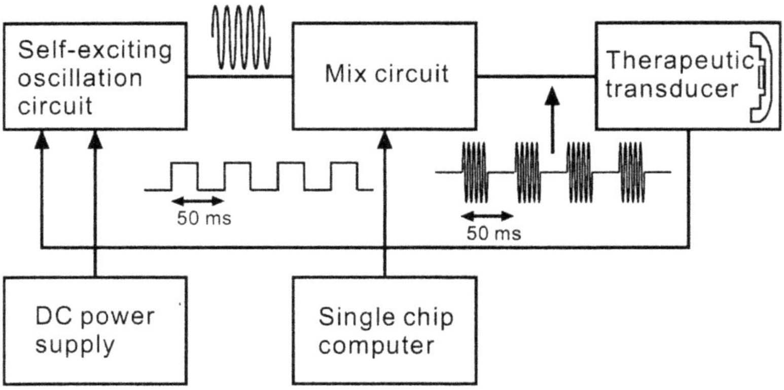

Figure 7 Block diagram of self-tuning power circuit on ultrasonic therapeutic transducers.

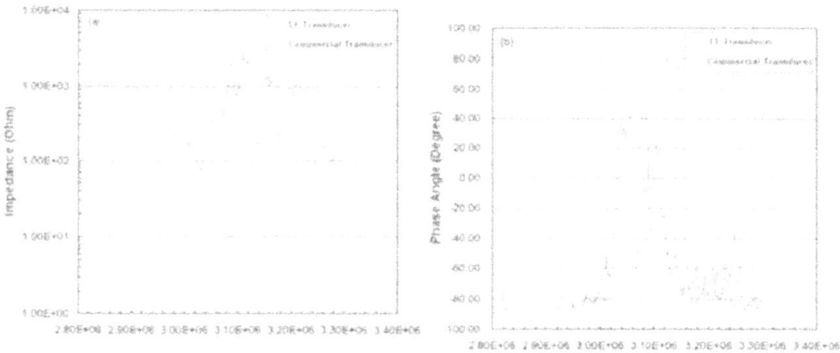

Figure 8 The plots of the impedance (a) and phase angle (b) of the LF transducer and commercial transducer versus the frequency.

DIRECT IN-SITU GROWTH OF MULTIWALLED CARBON NANOTUBES ON SILICON CARBIDE PARTICLES AS A PRECURSOR FOR FABRICATING SILICON CARBIDE – CARBON NANOTUBE COMPOSITES

Amit Datye, Kuang-Hsi Wu, George Gomes
Mechanical and Materials Engineering, Florida International University, Miami, FL 33174

Latha Kumari, Wenzhi Li
Department of Physics, Florida International University, Miami, FL 33199

Hua-Tay Lin
Oak Ridge National Laboratory, Oak Ridge, TN 37831-6068.

ABSTRACT

Carbon Nanotubes (CNTs) are 100 stronger than steel with only 1/6 of its weight and are ideal reinforcing fibers for nanocomposites. CNT has a thermal conductivity of 4,000 W/m-K, which is twice that of diamond – the best conductor on earth. CNTs can be either semi-conductor or metallic conductor, depending of its chirality. These remarkable properties enable CNTs to have enormous potential for advanced material applications, such as nanocomposites and in nanoelectronics. The application of CNTs on nanocomposites has been greatly hindered by two main factors – the interface bonding between the carbon nanotubes and the matrix material, and the difficulty in getting a uniform dispersion of carbon nanotubes in the matrix.

In the following research we present a new method of accomplishing uniform CNT distributions in a Silicon Carbide (SiC) matrix by direct in-situ growth of multiwalled nanotubes (MWNTs) on the SiC particles. This method ensures a more uniform dispersion of CNTs in the matrix than the traditional ex-situ CNT mixing methods. The other advantage of this method is that most of the CNTs grown are individual CNTs and not in the form of bundles.

INTRODUCTION

The increasing demand of ceramics as structural components has highlighted the utility of materials containing SiC (Silicon Carbide). SiC is a preferred ceramic material for many applications in harsh environmental conditions because of its resistance to high temperatures, aggressive chemicals and abrasion.[2] SiC is a covalent compound which can exist in the cubic or hexagonal crystal structures. SiC has excellent high-temperature strength because of this covalent nature and therefore, SiC holds promise as a heat-resistant material. At the same time, the covalent nature of SiC makes densification difficult[3]. SiC has been investigated as structural material in nuclear applications, such as in gas turbines, heat exchangers, and ceramic fans. Its high temperature strength, high thermal conductivity, and low nuclear activation, make it attractive for such applications.[4] In addition to interest in its properties for structural applications, SiC has more recently been the focus of many investigations for its electronic properties. As a wide-gap semiconductor, it has received considerable attention as a material for high-power device applications, including radar and microwave applications under extreme conditions. It has a high critical field strength, good carrier mobility, and excellent thermal conductivity.

Since the discovery of carbon nanotubes (CNTs) by Iijima[6] in 1991, they have been an attractive candidate for fundamental research. An elastic modulus greater than 1 TPa comparable to that of diamond (1.2 TPa) shown by theoretical and experimental results,[7-13] and with reported strengths 10–100 times higher than the strongest steel at a fraction of its weight makes carbon nanotubes an ideal candidate as reinforcing fibers in composites for structural applications. High thermal conductivity, twice as high as that of diamond[12-13] and stability in vacuum upto 2800°C make them ideal for thermal management applications. Therefore CNTs with their outstanding mechanical and thermal properties and extraordinarily high aspect ratios are the ideal reinforcing fibers for polymer and ceramic matrix composites.[14-16] Several applications were proposed for CNTs and many of which are concerned with conductive or high strength composites.[17-19] It is anticipated that the inclusion of CNTs in a ceramic matrix would allow one to produce composites with high stiffness and improved mechanical properties compared to that of the single phase ceramic material.[18]

Peigeny et al.[7, 28] and Balazsi et al. have tried ex-situ mixing of CNTs with Magnesium Aluminide and Silicon Nitride matrix, respectively, with limited success. The major problem that hinders the success of the CNT composite is uniform distribution of the CNTs in the matrix. Chemical functionalization of CNT surface has been one major effort for improving dispersion and nanotube/matrix interactions for processability and property enhancement. The surface modification will allow the unique properties of CNTs to be coupled with other matrix materials.[20-22] Several approaches to functionalization have been developed, including defect functionalization, covalent functionalization of the sidewalls, and nanocovalent exohederal functionalization with polymers or surfactants as supramolecular adducts,[23, 24] using CVD to roughen the surface and promote adhesion.[25] Yet another possibility is to expose the nanotube composite to gamma radiation for altering the chemistry at the interface of property enhancement, as has been done in SWNT-PMMA[26]. Plasma treatment of surface by Bubert et al.[27] was also attempted. Most of the previous research on SiC-CNT ceramic composites had been done by directly ex-situ mixing CNTs with SiC ceramic matrices. Various ex-situ techniques for uniform dispersion of CNTs in ceramic matrices have been tried but to limited success.[28, 29] SiC composites that contains 10 vol% MWNTs only had a 10% increase in bending strength and toughness over monolithic SiC.[28]

In this research we evaluate direct in-situ growth of CNTs on the SiC particles to circumvent the problem of poor dispersion of the CNTs inside the ceramic matrix.

MATERIALS AND METHODS

In-Situ Growth of MWNT's on SiC powders

A CVD growth process for CNTs involves heating the catalyst material typically transition metal nanoparticles on a support material to high temperature in a chamber and then exposing it to a hydrocarbon gas. The general nanotube growth mechanism in a CVD process includes the dissociation of hydrocarbon molecules catalyzed by the transition metal, and dissolution and saturation of carbon atoms in the metal nanoparticles. Then the carbon from the saturated metal particle precipitates, and this process leads to the formation of tubular carbon solids in sp^2 structure. For multi-walled carbon nanotube (MWNT) growth, ethylene (C_2H_4) or acetylene (C_2H_2) are mostly used as the carbon source in the CVD methods. The growth temperature is usually in the range of 550 - 750°C. Iron, nickel, or cobalt is often used as catalyst. The choice of using these transition metals as catalyst for CNT growth in CVD is based on the phase diagrams of the metals and carbon. Carbon has finite solubility in these metals at high temperatures, which leads to the formation of metal-carbon solutions and consequently the growth of the nanotubes. In the SiC system, which has higher sintering temperatures, cobalt is used in this study as a catalyst for the growth of nanotubes. Cobalt has been shown previously to give a higher weight percent yield of nanotubes. Cobalt(II) nitrate hexadydrate ($Co(NO_3){\cdot}6H_2O$, 98+%, A.C.S. reagent, Sigma-Aldrich), which is used as a catalyst precursor[30-32] and SiC powder (alpha Silicon Carbide – UK Abrasives with an average particle size of 500 nm) are mixed in ethanol, followed by sonication. Then the mixture is dried at 125°C in a furnace followed by grinding into fine powders.

SiC powders with different weight percent cobalt (Co/SiC = 1, 2, and 5 wt%) were placed in a tube furnace for the growth of CNTs on the SiC particles.

Removal of Catalyst and Purification of the Nanotubes

Sintering SiC requires high temperatures around 1950°C (with additives like Alumina, Magnesium Oxide) or 2250°C without additives for nearly (>95%) fully dense samples. SiC has been proved chemically unstable in the temperature range 1400-2000°K (1127°C - 1727°C) in the presence of cobalt, nickel, or iron. The instability is caused by the reaction of SiC with metals to produce silicides instead of carbides, leaving free carbon.[33] The SiC particles with in-situ grown MWNT's are therefore purified using acid treatment[34-36] to remove the cobalt catalyst and air oxidation treatment[37] to remove the amorphous carbon formed during the CVD growth process. The as grown SiC particles with CNTs are placed in a low vacuum filtration flask with 0.01 micron pore size polycarbonate filter papers and subjected to room temperature treatment with Hydrochloric acid[32] (HCl, 38%, Sigma Aldrich) for up to 4 hours to remove the catalyst particles. Since the MWNT's in this method can be formed by tip growth or root growth mechanism, the acid treatment can remove the catalyst particles revealed by tip growth whereas in the case of nanotubes with root growth the catalyst particles are immune from the acid attack by the graphitic layers surrounding it.

The SiC particles with in-situ MWNT's post acid treatment are then subjected to oxidation[37] at elevated temperatures (>400°C). This serves two purposes: to remove the amorphous carbon formed during the CVD process and to remove the graphitic layers surrounding the catalysts. Oxidation is followed by acid treatment again as previously mentioned to remove the exposed catalyst particles.

Characterization

The morphology of the MWNTs was examined by scanning electron microscopy (JEOL 6330F FEGSEM), and structural analysis was carried out using transmission electron microscopy (FEI CM20). Energy dispersive X-ray spectroscopy (EDS) attached to the SEM and X-ray examination was carried out on the samples before and after acid treatment to analyze the compositions. XRD was performed using a Siemens D5000 Diffraktometer employing Cu Ka radiation operated at 40 kV and 40 mA and fitted with a graphite monochromator. The step size and scan rate were fixed at 0.01° and 1 sec/step, respectively. The yield of the in-situ growth of CNTs on SiC mixtures was characterized using TGA (TA Instruments: High–Res TGA 2950 Thermogravimetric Analyzer). The samples were heated in an oxidizing atmosphere with a heating rate of 20°C/min in the temperature range 50–800°C under 20 ml/min dry air, and the weight change was recorded. Thermogravimetric analysis (TGA) was carried out to analyze the effect of selective oxidation on the particles, as well as to determine the weight percentage of the MWNTs. The specimens for TEM analysis were prepared by dispersing the samples in methanol and then spreading them on a TEM grid.

RESULTS AND DISCUSSION

Characterization of Initial Raw Powder

The initial raw powder was characterized using SEM and XRD. The SiC powder used in this study has an average powder size of 0.5 microns and the XRD patterns indicate a small amount of 6H phase present in the powder with a bulk of alpha phase SiC. Figure 1 below shows the FESEM image of the raw powder with an XRD scan showing the major alpha phase SiC.

Figure 1: FESEM and XRD of raw powder

SEM Characterization of CNT on SiC after growth

CNTs were grown in-situ on SiC particles with 1%, 2%, 5% and 20% weight percent catalyst. The 20% weight catalyst samples were chosen to verify the uniformity of CNT distribution on the ceramic particles at the macro level. Figures 2(a) and 2(b) below show CNTs grown on SiC particles with 1, 2 and 5 wt% Cobalt catalyst for 15 minutes growth time. SEM imaging shows that the CNTs are approximately 10-20 nm in diameters and approximately 2-5 microns in length for 15 minutes growth time. The most remarkable advantage of this in-situ growing CNT method is that, in addition to the uniformity of CNT distribution, most of the CNTs grown are individual CNTs. Neither bundle CNTs nor agglomerations of CNTs was noticed. As a consequence, the undesired agglomeration of CNTs, which is common in ex-situ mixing methods, is thus avoided. The CNTs grown by means of this method appear to wrap around the individual SiC particles nicely.

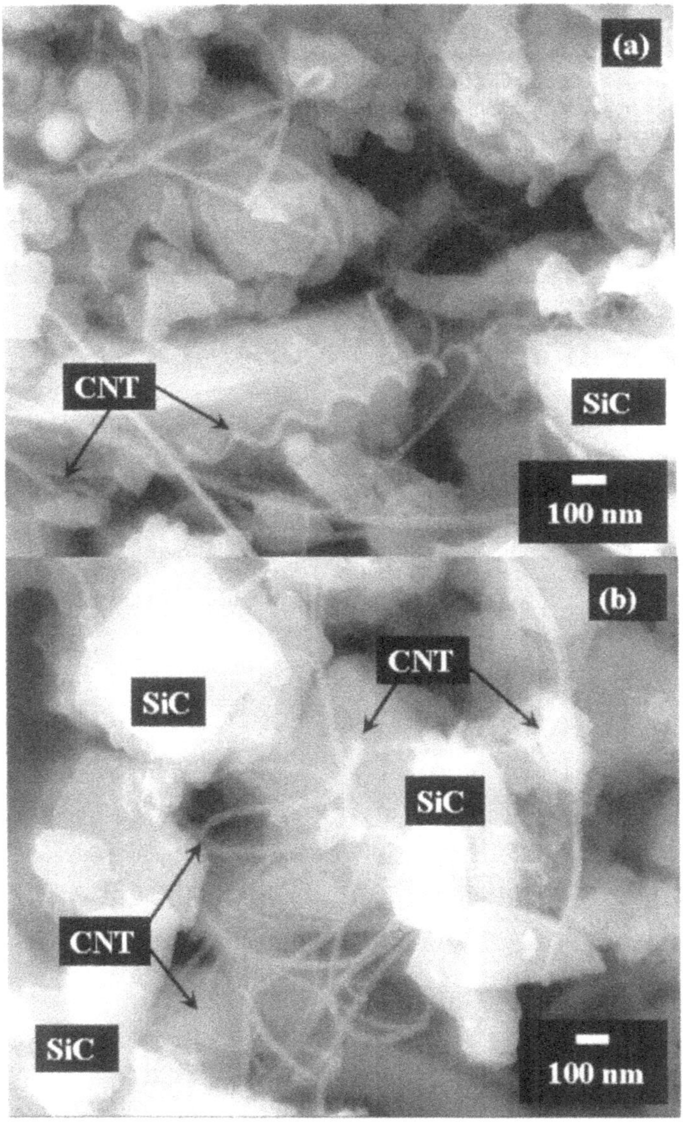

Figure 2: (a) and (b) CNT growth on SiC particles 1% wt and 5% wt. Co. - 15 min. growth

EDS Characterization after Catalyst Removal

EDS is used to qualitatively determine the reduction in cobalt after the acid treatment. It is evident from Figure 2 that starting from an initial 1 wt% of cobalt in the powder after the CNT growth there is there is approximately 50% reduction in the weight percent of cobalt after Step1 and another 25% of Co was further removed after Step 2.

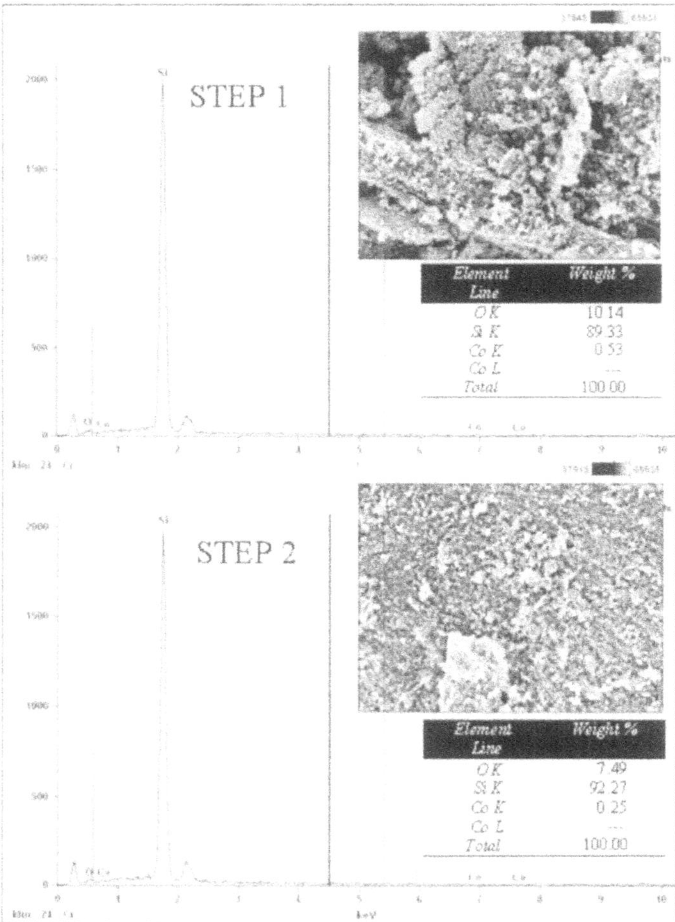

Figure 3: Qualitative EDS analysis to determine reduction in Cobalt.

TGA characterization of CNT yield post purification

TGA analysis was carried out by burning the sample with a heating rate of 20°C/min in the temperature range 50–800°C under 20 ml/min ambient air. Huang et al.[37] showed that MWNTs start decomposing above 400°C in an oxidizing atmosphere with a rapid loss at 600 °C after which the curve flattens [37].

The TGA curves (Figure 4) below show that there is very little amorphous carbon present in the mixture. It can be seen that there is a uniform trend of CNT yield directly proportional to the weight percent of cobalt in the mixture and the growth time. CNT yield for 1, 2 and 5 wt% Cobalt on the SiC particles is 1.75, ~2 and ~4%, respectively. It can also be seen that increasing the growth time to 30 minutes nearly doubles the CNT yield for all weight percentages of cobalt.

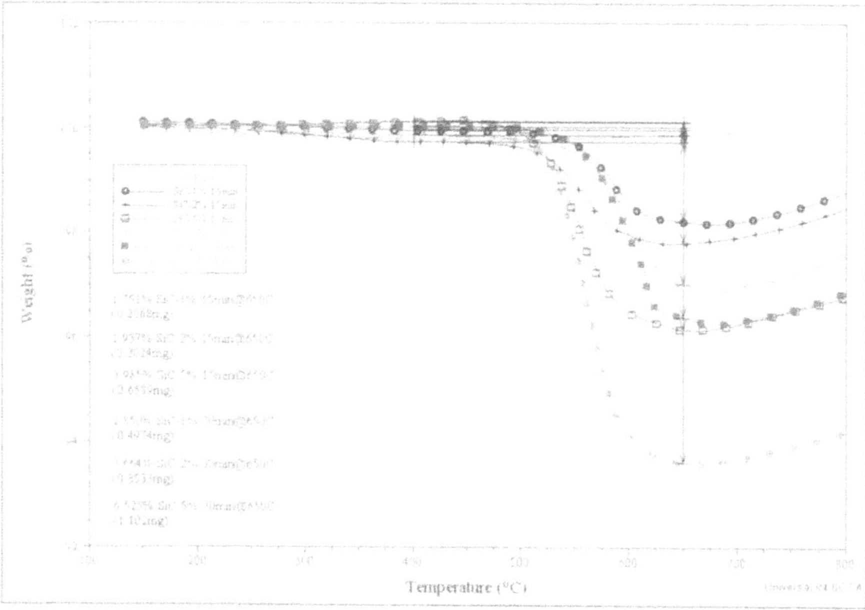

Figure 4: TGA Analysis of SiC with in-situ post purification

XRD Characterization of the powders with in-situ CNTs post purification

XRD patterns confirm that there is no formation of unwanted carbides of Cobalt during the CVD growth and there are no significant contaminations except some remaining Cobalt in the powders with in-situ CNTs after purification. Most of the peaks are from alpha phase SiC which corresponds to the initial raw powder composition. There is also a distinct carbon fullerite peak in all samples post purification indicating that the CNTs have survived the process.

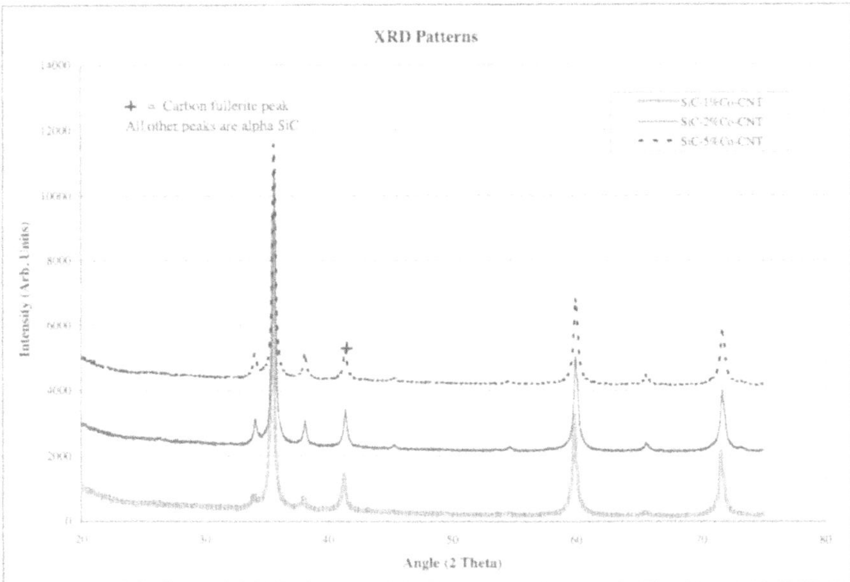

Figure 5: XRD analysis of the SiC with in-situ grown CNTs post purification and oxidation

SEM characterization of MWNT's on SiC post purification

SEM characterization post purification, as seen in Figures 6 (a) and (b) below for 1 wt% Cobalt and 5wt% cobalt, shows CNTs still present on the powders after purification by acid reflux and oxidation. It can be seen that the CNTs have not bundled or formed ropes as a result of the processing.

CNTs are also seen more or less uniformly distributed in the powder as seen in a low magnification image Figure 6(b).

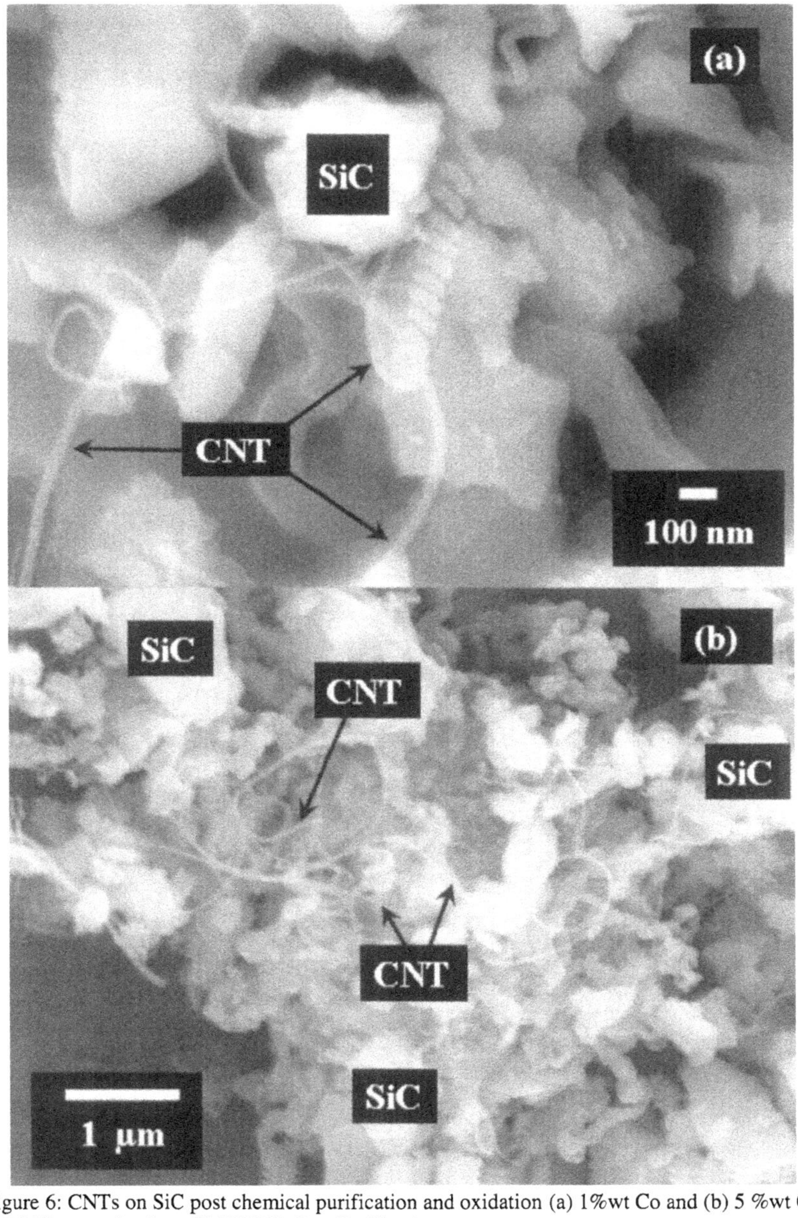

Figure 6: CNTs on SiC post chemical purification and oxidation (a) 1%wt Co and (b) 5 %wt Co

HRTEM Characterization of MWNT's Post Purification

HRTEM characterization post purification and oxidation of the CNTs grown in-situ on the SiC particles with 5 wt% Cobalt catalyst as seen below in Figure 7 show the presence of the CNTs on the SiC particles. It can also be seen from the HRTEM images that the percentage of the Cobalt catalyst particles has been significantly reduced.

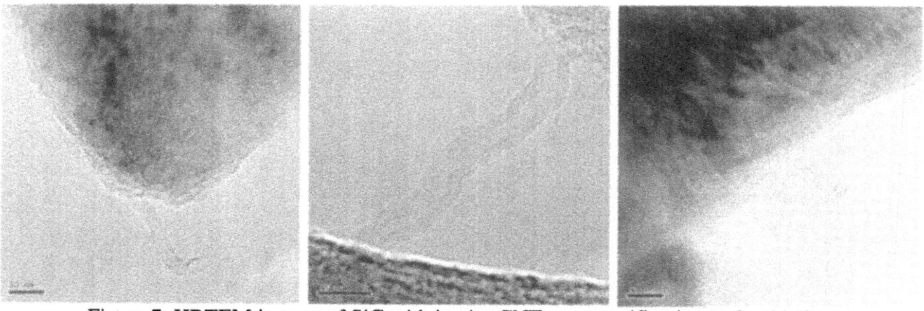

Figure 7: HRTEM images of SiC with in-situ CNTs post purification and oxidation

CONCLUSION

A method of directly growing MWNT's on SiC particles has been developed. This method has the following unique advantages.

1. This method can be used to manufacture Silicon Carbide-Carbon Nanotube composites with a more uniform dispersion of CNTs in the matrix than the traditional ex-situ mixing methods.
2. Most of the CNTs grown by this method are individual CNTs and not bundles of CNT found on the particles. Therefore these particles with CNTs grown in-situ are ideal for making CNT based composites.
3. CNTs grown for 30 minutes of growth time tend to wrap around the particles which is ideal for making composites.
4. In addition to the uniform dispersion throughout the matrix this method can be scaled up for large scale manufacturing and requires less processing time that traditional ex-situ CNT mixing.

ACKNOWLEDGEMENT

The authors would like acknowledge the support from the Office of Naval Research (ONR) Grant Number # N000140610131 and Dr. I. Perez of ONR for his support. L. Kumari, W.Z. Li acknowledge the support from Air Force Office of Scientific Research (AFOSR) Small Business Technology Transfer (STTR) Funding (Contract Numbers: FA9550-05-C-0126 for Phase I and FA9550-06-C-0136 for Phase II). L.Kumari and W. Z. Li acknowledge the support from NSF under grant DMR-0548061.

REFERENCES

1. R. Ruh and A. Zangvil, "Composition and Properties of Hot-Pressed SiC-AlN Solid Solutions," in 82nd Annual Meeting, The American Ceramic Society.
2. A. Can, M. Herrmann, D. S. McLachlan, I. Sigalas, and J. Adler, "Densification of liquid phase sintered silicon carbide," Journal of the European Ceramic Society 26 1707-13 (2006).
3. Y. Shinoda, M. Yoshida, T. Akatsu, and F. Wakai, "Effect of Amount of Boron Doping on Compression Deformation of Fine-Grained Silicon Carbide at Elevated Temperature," J. Am. Ceram. Soc., 87[8] 1525-29 (2004).
4. T. Yamamoto, H. Kitaura, Y. Kodera, T. Ishii, M. Ohyanagi, and Z. A. Munir, "Consolidation of Nanostructured beta-SiC by Spark Plasma Sintering," J. Am. Ceram. Soc., 87[8] 1436–41 (2004).
5. Y. Pan, "Chemical Instability of Silicon Carbide in the presence of transition metals," J. Am. Ceram. Soc., 79[8] 2017-26 (1996).
6. Iijima S. Helical microtubules of graphitic carbon. Nature 1991; 354:56–8.
7. A. Peigney, Ch. Laurent, E. Flahaut, and A. Rousset Carbon nanotubes in novel ceramic matrix nanocomposites. Ceram Int 2000; 26:677–83.
8. G. Van Lier, C. Van Alsenoy, V. Van Doren, and P. Geerlings, Ab initio study of the elastic properties of single-walled carbon nanotubes and graphene. Chem Phys Lett 2000; 326:181–5.
9. M. Treacy, T.W. Ebbesen, and J.M. Gibson, Exceptionally high Young's modulus observed for individual carbon nanotubes. Nature 1996; 381:678–80.
10. M.F. Yu, O. Lourie, M.J. Dyer, K. Moloni, T.F. Kelly, and R.S. Ruoff, Strength and breaking mechanism of multiwalled carbon nanotubes under tensile load. Science 2000; 287:637–40.
11. A. Thess, R. Lee, P. Nikolaev, H. Dai, P. Petit, J. Robert, C. Xu, Y.H. Lee, S.G. Kim, A.G. Rinzler, D.T. Colbert, G.E. Scuseria, D. Tomanek, J.E. Fischer and R.E. Smalley, Crystalline ropes of metallic carbon nanotubes. Science 1996; 273:483–7.
12. Ando Y, Zhaoa X, Shimoyama H, Sakai G, and Kaneto K, Physical properties of multiwalled carbon nanotubes. Int J Inorg Mater 1999; 1:77–82.
13. M.J. Biercuk, M.C. Llaguno, M. Radosavlijevic, J.K. Hyun, and A.T. Johnson, Carbon nanotube composites for thermal management. Appl Phys Lett 2002; 80:2767–9.
14. R.H. Baughman, A.A. Zakhidov, and W.A. de Heer, Carbon nanotubes—the route toward applications. Science 2000; 297:787–92.
15. K.M. Prewo, Fiber-reinforced ceramics: new opportunities for composite materials. Am. Ceram Soc Bull 1989; 68:395–400.
16. Breuer and Uttandaraman Sundararaj, Polymer Composites, Dec. 2004, Vol. 25, No. 6, p. 630-645.
17. E.T. Thostenson, Z. Ren, and T.W. Chou, Advances in the science and technology of carbon nanotubes and their composites: a review. Compos Sci Technol 2001; 61:1899–912.
18. H. Dai, Carbon nanotubes: opportunities and challenges. Surf Sci 2002; 500:218–41.
19. F. Lupo, R. Kamalakaran, C. Scheu, N. Grobert, and M. Ruhle, Microstructural investigations on zirconium oxide—carbon nanotube composites synthesized by hydrothermal crystallization. Carbon 2004; 42:1995–9.
20. Min-Feng Yu, J. of Eng. Materials and Technology, July 2004, Vol. 126, p. 271-278.
21. Deepak Srivastava, Chenyu Wei, Kyeongjae Cho, Appl. Mech. Rev., Vol. 56, No. 2, March 2003, p. 215-230.
22. T. Tiano, M. Roylance, and J. Gassner, 32nd International SAMPE Technical conference, 192, 2000.
23. K. Lozano, "Vapor-grown carbon-fiber composites: Processing and electrostatic dissipative applications", JOM, Vol. 52, issue 11, pp. 34-36, November 2000.
24. A. Star, J. Stoddart, D. Steuerman, M. Diehl, A. Boukai, E. Wong, X. Yang, S. Chung, H. Choi, J. Heath, Preparation and Properties of Polymer-Wrapped Single-Walled Carbon Nanotubes, Angewandte Chemie International Edition, 2001, 40, 1721–1725.

25. X. Gong, J. Liu, S. Baskaran, R. D. Voise, and J. S. Young, Surfactant-assisted processing of carbon nanotube/polymer composites, Chem. Mater. v. 12, 1049-1052 (2000).
26. Y. Lin, A. M. Rao, B. Sadanadan, E. A. Kenik, and Y. Sun, Functionalizing Multiple-Walled Carbon Nanotubes with Aminopolymers, J. Phys, Chem, B., 106,p. 1294, (2002)
27. Harmon, J P; Muisener, P A O; Clayton, L; D'Angelo, J; Sikder, A K; Kumar, A; Meyyappan, M; Cassell, A M Surface Engineering 2001--Fundamentals and Applications (as held at the 2001 MRS Fall Meeting); Boston, Massachusetts; USA; 26-29 Nov. 2001. pp. 425-435. 2002
28. Peigney, Ch. Laurent, E. Flahaut, and A. Rousset Carbon nanotubes in novel ceramic matrix nanocomposites. Ceram Int 2000; 26:677–83.
29. Ma, R.Z., Wu, J., Wei, B.Q., Liang, J., and Wu, D.H., Processing and properties of Carbon Nanotube/nano-SiC ceramic, J. Mater. Sci. 33, 5243 (1998).
30. D. Bethune, C. Klang, M. De Vries, G. Gorman, R. Savoy, J. Vazquez, and R. Beyers, Cobalt-catalysed growth of carbon nanotubes with single-atomic-layer walls, Nature 363, 605 (1993).
31. H.J. Dai, A.G. Rinzler, P. Nikolaev, A. Thess, D.T. Colbert, and R.E. Smalley, Single-wall nanotubes produced by metal-catalyzed disproportionation of carbon monoxide. Chem. Phys. Lett 1996; 260: 471.
32. W.Z. Li, J.G. Wen, M. Sennett, and Z.F. Ren, Clean double-walled carbon nanotubes synthesized by CVD. Chem Phys Lett 2003; 368:299–306.
33. Y. Pan, Chemical Instability of Silicon Carbide in the presence of transition metals. J. Am. Ceram. Soc. 79, 2017-2026 (1996).
34. F. Li, H. Cheng, Y. Xing, P. Tan and G. Su, Purification of single walled carbon nanotubes synthesized by the catalytic decomposition of hydrocarbons. Carbon 2000; 38: 2041–5.
35. P. Umek, D. Vrbanic, M. Remskar, T. Mertelj, P. Venturini, and S. Pejovnik, An effective surfactant-free isolation procedure for single-wall carbon nanotubes. Carbon 2002;40:2581–5.
36. K. Strong, D. Anderson, K. Lafdi, and J. Kuhn, Purification process for single-wall carbon nanotubes. Carbon 2003; 41:1477–88.
37. Z. Shi, Y. Lian, F. Liao, X. Zhou, Z. Gu and Y. Zhang, Purification of single-wall carbon nanotubes. Solid State Commun 1999;112:35–7.
38. W. Huang, Y. Wang, G. Luo, and F. Wei, 99.9% purity multi-walled carbon nanotubes by vacuum high-temperature annealing, Carbon 41 (2003) 2585–2590.
39. Biercuk, M.J., Llaguno, M.C., Radosavlijevic, M., Hyun, J.K., and Johnson, A.T., Appl. Phys. Lett. 80, 2767 (2002).

Photocatalytic Materials

EFFECT OF FUNCTIONAL COATINGS ON TOPOGRAPHICAL PROPERTIES OF GLASS

Minna Piispanen[a], Mikko Aromaa[b], Jyrki M. Mäkelä[b] and Leena Hupa[a]
[a]Åbo Akademi University, Process Chemistry Centre, Turku, Finland
[b] Tampere University of Technology, Department of Physics, Tampere, Finland

ABSTRACT

In this work the effect of two photocatalytic coatings on surface topography of float glass is discussed. The goal was to compare different instrumental methods and parameters for achieving a reliable description of surface topography. One coating was a sol-gel derived titania and the other was a titania-silver coating applied by liquid flame spray method. The titania coating should gave a smooth covering of the glass surface, while the titania-silver surface consisted of nano-to-micro particles. Surfaces were imaged with SEM. The roughness values were determined by white light confocal microscopy (COM) and atomic force microscopy (AFM). COM was utilized for 3D surface mapping and for determining average topographic parameters. AFM was used to characterize the height asymmetry, and number, size and form of local maxima of the surfaces. The surface energy was determined by measuring contact angles of three liquids of known properties. The results indicated that roughness correlated with wettability but also that the photocatalytic character of the surfaces should be considered.

INTRODUCTION

Detailed characterization of surface microstructure has become an important criterion of surfaces[1-4]. Topographic characterization is utilized in estimating the influence of additional films on the roughness of a surface. Additional films are applied for functionalizing the surfaces for e.g. better corrosion and wear resistances, or soil repellence and cleanability. Self-cleaning and/or easy-to-clean coatings have been employed to enhance the cleanability of glass and ceramics[5-9]. The principles of self-cleaning surfaces have been studied widely[5, 6, 9, 10]. By self-cleaning or easy-to-clean coatings, less and milder cleaning agents can be used, thus giving more economical and environmental products.

An enhanced cleanability can be achieved by implementing either a hydrophobic or a hydrophilic film on the surface. The coatings studied in this work were hydrophilic. Photocatalytic titania coatings which utilize sunlight and rinsing water are the best known examples of surfaces with hydrophilic properties. Titania has been applied to the surface by different methods, such as flame spraying, atomic layer deposition and sol-gel procedures[11-13]. Functionality of the coatings has been reported widely[11-15]. In some reports the chemical and mechanical wear of additional coatings has been discussed[17-19].

In this work the effect of additional coatings on nano-, micro- and macro-roughnesses were studied. The goal was to specify which measuring methods and parameters give a reliable three-dimensional description of the topography of non-ideal surfaces. The surface topography was characterized with scanning electron microscopy (SEM), confocal optical microscopy (COM) and atomic force microscopy (AFM). SEM shows a two dimensional topographic image of the surface. COM and AFM give a three dimensional characterization and topographical parameters for specifying the surface. These parameters enable a true three-dimensional quantification of the surface. The surface parameters enable a better understanding of the influence of topography on surface wetting, and thus soil attachment and cleanability.

EXPERIMENTAL

Surface topography of float glass coated with experimental sol-gel derived titania and liquid flame sprayed titania-silver was studied by different instrumental methods. Both coatings were manufactured from tetraisopropyl orthotitanate. The manufacturing parameters have been described in detail elsewhere[12,13,14,18,20]. The hydrophilic and photocatalytic sol-gel derived titania surface was manufactured by dip-coating the glass substrates in the sol and withdrawing at a speed of 0.47 mm/s. To obtain a ceramic TiO_2 coating, the sol-gel surface was heat-treated at a rate of 1 °C/min to 500°C and maintained at 500°C for 1 h and finally allowed to cool in the furnace. The hydrophilic, photocatalytic and antibacterial titania-silver coating was manufactured with the Liquid Flame Spray (LFS) method in a laboratory-scale coating unit[11,12,20]. Before spraying with titania-silver nanoparticles the glass plates were preheated to 500-800°C and finally annealed in order to avoid surface cracking. The nanoparticles not attached to the surface were wiped off with isopropanol.

The surface topography for the reference uncoated and the experimental coated surfaces was studied with scanning electron microscopy, white light confocal microscopy and atomic force microscopy. The scanning electron microscope (FEG-SEM-EDXA, LEO 1530 from Zeiss) was mainly used to characterize the overall surface appearance and to give the chemical composition of the surface. The thickness of the coatings was measured by XPS (X-ray Photoelectron Spectroscopy, PHI Quantum 2000). A spinning disk confocal microscope with a white light source (COM, NanoFocus µSurf®) was used to determine 3D surface parameters. The average 3D surface roughness values (S_a and S_q) according to DIN EN ISO 4287 were measured at 100x and 20x magnifications. At 100x magnification the vertical resolution of the lens is 2 nm and the numerical aperture 0.8-0.95 for a measurement area of 160-158 µm. At 20x magnification the vertical resolution is 6 nm and the numerical aperture 0.40-0.46 for a measurement area of 800-772 µm. For the measurements done with the 100x lens the cut-off wavelength of 80 µm was used. The cut-off wavelength for the 20x measurements was 250 µm. The average roughness value, S_a, is the arithmetic mean of absolute values of the surface departures from the mean plane. S_q gives the root mean square roughness (RMS), obtained by squaring each height value in the dataset, then taking the square root of the mean. The roughness values were measured for 6 parallel samples.

Atomic force microscopy (AFM, Nanosurf® Mobile S) was performed in the dynamic operating mode using silicon cantilevers with a resonance frequency of 190 kHz. The maximum scan size was 110 µm with a maximum vertical range of 22 µm. The vertical resolution was 0.34 nm and the horizontal resolution 1.7 nm. Size of the analyzed areas was 5x5 µm. AFM measurements of the surfaces were performed with a speed of 1 line/s. Image processing was done using 512x512 pixel frame grabber in air without filtering. Besides the average roughness values S_a and S_q, six additional parameters (S_p, S_v, S_t, S_{sk}, S_{ku}, S_z) were used to characterize the surface topography. These parameters describe the amplitude properties, which give information about the statistical average properties, the shape of the height distribution histogram and extreme topographical variations. S_p and S_v are by definition the maximum profile peak height and the maximum profile valley depth, while S_t gives the distances from the mean line to the highest/lowest point in the evaluation length/area. S_{sk} is a skewness parameter and measures the asymmetry of the profile about the mean plane. A negative skew indicates a predominance of valleys, while a positive skew is seen on peak rich surfaces. S_{ku}, kurtosis is a measure of sharpness of the surface, or the distribution of spikes above and below the mean line. For spiky surfaces $S_{ku} > 3$, for bumpy surfaces $S_{ku} < 3$ and perfectly random surfaces have kurtosis 3. S_z is the ten point average of the absolute heights of the five highest peaks and five deepest valleys (EUR15178N report). A thorough description of the roughness parameters can be found elsewhere[1-3]. With AFM the topographical parameters are given as an average value of three parallel samples.

The surface wettability was determined by measuring the static water contact angle for an ultra pure water drop for 20 seconds collecting one image per second by a contact angle meter (KSV CAM200) at room temperature. The contact angle was calculated using the Young-Laplace equation.

Also the surface energy was calculated from the contact angles of three liquids of known properties (water, ethylene glycol, diiodmethane). The surface energy was measured for 5 parallel samples.

RESULTS AND DISCUSSION

The visual appearances of the uncoated and coated glass surfaces did not differ significantly. However, the titania-silver surfaces had a yellowish tint arising from the silver in the coating. The SEM micrographs of the reference glass and the sol-gel derived titania surface did not show marked topographical differences (Figure 1). However, in the titania surface a crack was seen. It was assumed the coating was either too thick or its thermal expansion was incompatible with the substrate. The EDX analysis of the sol-gel surface indicated that the glass was totally covered by titania. According to XPS the thickness of the coating was 350 nm. Thin-Film X-Ray Diffraction of the titania coating manufactured with the same parameters as in this work has been reported to have anatase crystal structure[13]. The SEM image of the liquid film sprayed titania-silver surface showed nanoparticles and up to 3-4 μm agglomerates of titania (Figure 1). According to XPS the coating thickness was only 2-3 nm. The same titania-silver coating as used in this work has been reported to give both anatase and rutile crystal structure; the small particles and agglomerates were anatase, while the larger agglomerates were rutile[21]. The surface also showed some open pore formation.

Figure 1: SEM micrographs of glass surfaces; (A) uncoated reference, (B) sol-gel derived titania coating and (C) LFS titania – silver coating showing a titania agglomerate, Ti

The average roughness values S_a and S_q measured by COM are given in Table 1. For all three surfaces the S_a values were lower than the S_q values. Roughness values measured with the lower magnification (20x) were higher than the values measured with the larger magnification (100x). A greater number and variety of peaks and valleys were taken into account when analyzing larger areas. Furthermore, surface flaws formed during e.g. washing and sample preparation were more abundant in the analysis of the larger areas. The larger magnification used for the small areas described the influence of local variations of the surface. If properly optimized the cut-off wavelength allows one to eliminate the effect of the distortion of the glass surface on the final roughness values. The lower magnification was assumed to give macro-scale roughness values including surface flaws, while the higher magnification gave a detailed description of micro- and nano-scale surface variations.

The S_a and S_q values were higher for the titania-silver surface than for the reference and titania coated surfaces (Table 1 and Figure 1). The higher roughness values originated from the titania particles and agglomerates on the surface. Similar differences in the roughness parameters S_a and S_q were obtained when analyzing with the 20x lens. The reference glass and the sol-gel derived titania surfaces had a similar S_a value, while the S_q value of the titania surface was clearly higher at low magnification. The COM images taken at lower magnification of the larger surfaces included cracks and manufacturing flaws affecting the roughness. Only crack free areas were imaged with the 100x lens. Basically, crack formation in the sol-gel derived titania coating should be avoided by optimizing the coating process.

Table 1: The average roughness value, S_a, and root mean square roughness (RMS), S_q, for the experimental surfaces measured with COM using lenses giving 100x and 20x magnifications.

	ref **[nm]**		**TiO_2** **[nm]**		**TiO_2-Ag** **[nm]**	
	Sa	Sq	Sa	Sq	Sa	Sq
glass (100x)	4 ± 1	6 ± 1	3 ± 1	5 ± 2	7 ± 2	13 ± 9
glass (20x)	22 ± 1	29 ± 1	20 ± 2	58 ± 10	31 ± 4	54 ± 10

The data given by the two lenses suggested that COM typically gives overall roughness values of real surfaces, especially for the lower magnification. In contrast to COM, AFM enables an atomic level resolution of the surface. A typical analysis area with AFM is a few micrometers only. Due to the high resolution, a micro-scale roughness of the analyzed area is obtained. Roughness parameters and images of the experimental surfaces obtained by AFM analysis are given in Table 2 and in Figure 2.

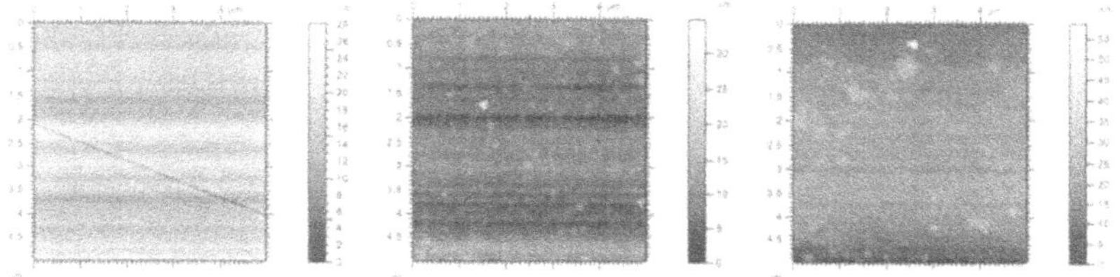

Figure 2: AFM images of the reference glass (left), sol-gel derived titania coating (middle) and liquid flame sprayed titania-silver coating (right). Image size 5x5 μm

The AFM images (Figure 2) show that the reference glass surface was smooth but has some nano scratches, probably originating from the cleaning stage. The scratches increased the roughness but in general the surface showed low topographic variation. The sol-gel derived titania surface had a homogenous nano structure. On the LFS titania-silver surface, particles and agglomerates are clearly seen. Only small variations in surface roughness can be measured with AFM. Thus, on non-ideal surfaces containing scratches, particles and/or bubbles a large amount of images should be taken to get a good description of the average roughness. AFM analysis is laborious because every disturbing factor will be taken into account. The roughness parameters in Table 2 show that the reference glass surface had the lowest S_a value, while the coated surfaces had higher S_a values. AFM gave lower S_a values than COM for the reference surface and the titania-silver coating, while relatively similar values were obtained with both analytical methods for the sol-gel derived titania surface. This suggests that the morphology of the titania coating was homogeneous when studying areas with no manufacturing flaws such as deep cracks. S_a and S_q values measured by AFM follow the same trend as seen for COM; S_q was higher than S_a depending on the different calculation principle. Large differences in S_a and S_q would indicate presence of large departures from the mean plane. For the sol-gel derived titania surface S_q measured by AFM and COM gave similar values, thus confirming the homogeneous roughness of the measured surface areas.

Further information on the surface structure can be deduced from the S_p and S_v values, i.e. the maximum peak height and valley depth compared to the mean surface line. The titania-silver surface had the highest valleys and peaks, while values were lower for the reference surface. Accordingly, the maximum vertical distance between the highest and lowest points, S_t was largest for the titania-silver surface and lowest for the reference glass. The asymmetry in the surface profile about the surface mean plane, S_{sk} suggested a negative skew, i.e. predominantly valleys for the reference surface probably due to scratches (Figure 2). Both coated surfaces had a positive skew and therefore a predominance of peaks. However, the S_{sk} value for the sol-gel derived titania surface was close to zero and was therefore close to a Gaussian like surface. The sharpness of the surface spikes can be estimated by the kurtosis value, S_{ku}. The reference and titania-silver surfaces had kurtosis >3, thus indicating spiky surfaces. The titania surface had a kurtosis value close to 3, thus suggesting a perfectly random surface with a Gaussian like distribution of valleys and peaks. Finally, the ten point average of the absolute heights of the five highest peaks and five deepest valleys, S_z, was highest for the titania-silver surface and lowest for the reference glass. For the surfaces studied, S_z did not give any additional information. The results suggest that AFM can be utilized to give an accurate description on local conditions of real surfaces when analyzing areas with no surface flaws. Large deviations from the mean surface level can be difficult to measure accurately. Using only one parameter, e.g. S_a, is not sufficient to give a reliable description of the surface. Surface parameters such as S_{sk} and S_{ku} describing the surface profile and sharpness could be included in the surface topography analysis. The parameters discussed in this work are amplitude parameters. In order to understand the surface topography better, hybrid parameters should be included in the analysis. These parameters reflect slope gradients and their calculations are based on local z-slopes. The software used in this work did not allow hybrid parameters.

Table 2: Average roughness value (S_a) and root mean square roughness (RMS, S_q), maximum profile peak height (S_p), maximum profile valley depth (S_v), the sum of S_p and S_v (S_t), skewness (S_{sk}), kurtosis (S_{ku}) and ten point average of the absolute heights of the five highest peaks and five deepest valleys (S_z)

	glass	glass + TiO2	glass + TiO2-Ag
S_a [nm]	1.4 ±0.3	3.0 ±1.2	3.0 ± 1.0
S_q [nm]	1.9 ±0.6	3.8 ±1.5	4.0 ± 1.1
S_p [nm]	8.7 ±1.6	23.2 ±2.5	33.4 ± 8.1
S_v [nm]	12.9 ±5.5	14.0 ± 6.0	17 ± 3.9
S_t [nm]	21.6 ± 7.1	37.3 ±3.5	50.3 ± 10.8
S_{sk}	-0.8 ± 0.7	0.1 ± 0.3	0.1 ± 0.7
S_{ku}	6.6 ±3.4	3.7 ±1.0	6.4 ± 2.8
S_z [nm]	18.7 ± 6.7	29.5 ±10.6	31.9 ± 2.7

The wetting properties of the surfaces were studied by measuring the contact angles of water and by determining the surface energy. The surface roughness has an impact on wetting properties[22,24]., A hybrid roughness parameter, S_{dr} can be used to calculate the r-value in the well-known Wenzel equation[22]. The S_{dr} is the ratio of the increment of the interfacial area of a surface over the sampling area. This parameter is often used to describe the "complexity" of the surface. A perfectly smooth surface (no height deviations) would have an S_{dr} of 0%. According to Wenzel, the relation between the roughness dependent measured contact angle θ_m and Young's contact angle θ_Y corresponding to an ideally flat surface may be written as

$$\cos\theta_m = r\cos\theta_Y \tag{1}$$

where r denotes the ratio between the real and projected surface areas of the samples. The equation states that a rough hydrophilic surface ($\theta_Y < 90°$) should be more hydrophilic, and a rough hydrophobic surface ($\theta_Y > 90°$) more hydrophobic than a flat surface with the same chemical composition. The roughness parameter S_{dr} may be used to calculate r from

$$r = 1 + S_{dr} / 100 \tag{2}$$

On similar surfaces as used in this work an r-value of approximately 1 has been reported[3]. This would mean that the roughness only marginally increases the surface area. Thus, the Wenzel equation has not to be taken into account in the contact angle studies for these surfaces.

Van Oss et al[23] proposed that the surface free energy can be determined by summing the Liftshitz - van der Waals component (dispersion component), α_{SV}^{LW}, and the Lewis acid-base components, α_{SV}^{-} and α_{SV}^{+} (polar components). The surface energy components can be calculated from the contact angle values of three liquids with known properties (water, ethylene glycol and diiodmethane) according to Equation 3[23]

$$(1+\cos\theta_{SL})\gamma_{LV} = 2\left(\sqrt{\gamma_{LV}^{LW}\sigma_{SV}^{LW}} + \sqrt{\gamma_{LV}^{+}\sigma_{SV}^{-}} + \sqrt{\gamma_{LV}^{-}\sigma_{SV}^{+}}\right) \tag{3}$$

The total surface free energy of the solid is then given by:

$$\sigma_{SV}^{tot} = \sigma_{SV}^{LW} + \sigma_{SV}^{AB} = \sigma_{SV}^{LW} + 2\sqrt{\sigma_{SV}^{+}\sigma_{SV}^{-}} \tag{4}$$

This approach was used to calculate the surface free energies for the experimental surfaces. The titania-silver and reference surfaces showed the same surface energy 37 mN/m, while the value for the titania coating was 45 mN/m (Table 3). The titania-silver coating was very thin (2-3 nm), thus partially explaining the surface energy value. Kanta et al[26] illustrated that high energy surfaces are wetted by most liquids because of their higher polarizability rather than simply because of their high surface energy. The contact angle values of the titania and titania-silver surfaces were similar, while the contact angle of the reference glass was higher. The titania and the titania-silver surfaces were more polar (higher acid-base values) than the reference surface, thus explaining the lower water contact angles. The wetting, one of the most important surface chemical properties, is controlled by the surface OH group density[24-25]. The surface OH groups can work as effective adsorptive and reactive sites. Thus, the surface OH groups are considered to be a major factor governing the chemical properties of surface films. Furthermore, the amount of the carbon substances adsorbed from the atmosphere is dependent on the surface OH group density of the films[24-25]. When surfaces adsorb carbon they get more hydrophobic. It has been reported that when a clean hydrophilic silica surface (water contact angle 0°) was exposed to open atmosphere, it gradually turned into mildly hydrophobic due to organic contamination[24-27]. In this work the water contact angle of the uncoated surface was higher than 60°. This suggests that the reference surface adsorbed carbon, thus getting more hydrophobic with time, when exposed to open atmosphere. The contact angles of the titania and titania-silver surfaces were about 30° lower than the contact angle of the glass surface. It was assumed that the titania and titania-silver surfaces partly decomposed the organic compounds adhered to them with the help of natural ligh. Accordingly, the surfaces were less hydrophobic than the reference surface. This could be taken as an indication of the photocatalytic character of the titania and titania-silver coatings. According to AFM, both titania coated surfaces had a slightly higher roughness than the reference surface. The relative trend in the differences of the water contact angles and average roughnesses was observed to be similar. Thus, the contact angle was assumed to depend both on the photocatalytic character and the nano roughness of the surfaces. Which property has a greater affect on the contact angle could not be verified with these experiments only. The photocatalytic character of the coatings in UV-light exposure has been reported elsewhere[11-13].

Table 3: Contact angle values and surface free energies of uncoated and titania coated glasses

	Acid-Base				Water
	LW	A	B	surface free energy [mN/m]	contact angle
Ref	35.9 ± 0.6	0.2 ± 0.2	4.5 ± 0.2	37.4 ± 0.8	65.1 ± 2.4
TiO_2	41.1 ± 0.3	0.3 ± 0.1	6.7 ± 0.1	45.4 ± 0.8	36.9 ± 0.8
TiO_2-Ag	31.6 ± 1.6	0.4 ± 0.2	7.3 ± 0.3	37.3 ± 1.5	37.6 ± 3.4

CONCLUSIONS

The effect of a sol-gel derived ceramic titania coating and a liquid flame sprayed titania-silver coating on the surface topography of float glass was measured with COM and AFM. The roughness parameters were measured for different sample sizes depending on the instrument. The surface roughness values obtained were found to be characteristic of the image size. Roughness analysis is sensitive to the measuring parameters. When measuring real surfaces, AFM is suited to give a detailed description on rather limited areas with no large surface flaws nor variations from the mean surface. Using only the average roughness values, S_a and S_q, does not give an unambiguous description of the surface topography. However, taking into account surface parameters such as S_{sk} and S_{ku} is likely to give a better characterization of the topography. These parameters describe the variation from a Gaussian surface. While AFM probed the nano- and micro-roughness of the surfaces, COM could be utilized to obtain an overall roughness of the surface so that large surface variations and topographical flaws were included. Thus, a micro- and macro-scale topography description was possible with COM. When comparing roughness parameter data for non-ideal surfaces, a thorough understanding of the limitations and variables of the method used is required.

The roughness values indicated that the reference glass surface was smooth but not a Gaussian like, due to some nano scratch in the surface. The sol-gel derived titania coating showed Gaussian behavior, even though it contained some homogeneous irregularities. The LFS titania-silver coating consisted of nanoparticles and some larger particle agglomerations on the surface, affecting the roughness parameters and giving the roughest surface.

Surface roughness is known to affect the contact angle. Although there was a correlation between roughness values and water contact angle, the role of the photocatalytic activity remains unclear. Due to the variations in the roughness values obtained with COM, probing the macroscale and with AFM, probing the nano to microscale, it was thought that the photocatalytic property could have a more significant effect on the contact angle.

ACKNOWLEDGEMENTS

This work is part of the activities of the Åbo Akademi Process Chemistry Centre funded by the Academy of Finland in their Centres of Excellence Program. Funding has been provided also by the Graduate School of Materials Research. Beneq Oy is acknowledged for the collaboration.

REFERENCES

[1]Stout, K.J., Development of Methods for the Characterisation of Roughness in Three Dimensions, ISBN 1-8571-8023-2

[2]J. Peltonen, M. Järn, S. Areva, M. Linden, J.B. Rosenholm, Topographical Parameters for Specifying a Three-Dimensional Surface, *Langmuir*, **20** (22), 9428-9431 (2004)

[3]M. Järn, S. Areva, V. Pore, J. Peltonen, M. Linden, Topography and Surface Energy Dependent Calcium Phosphate Formation on Sol-Gel Derived TiO_2 coatings, *Langmuir*, **22**, 8209-8213 (2006)

[4]L. Fröberg, L. Hupa, Topographic Characterization of Glazed Surfaces, *Appl. Surf. Sci.*, **254**(6),1622-1629 (2008)
[5]A. Fujishima, T.N. Rao, D.A. Tryk: Titanium Dioxide Photocatalysis. *J. Photochem. Photobiol. C: Photochem.* Rev., **1,** 1-21(2000).
[6]A. Fujishima, Z. Zhang, Titanium Dioxide Photocatalysis: Present Situation and Future Approaches. *C.R.Chemie*, **9,** 750-760 (2006).
[7]R. Wang, K. Hashimoto, A. Fujishima, M. Chikuni, E. Kojima, A. Kitamura, M. Shimohigoshi, T. Watanabe: Photo-Generation of Highly Amphiphilic TiO_2 Surfaces. *Adv. Mater.*, **10,** 135-138 (1998).
[8]A. Nakajima, K. Hashimoto, T. Watanabe: Transparent Superhydrophobic Thin Films with Self-Cleaning Properties. *Langmuir*, **16,** 7044-7047 (2000).
[9]X.M. Li, D. Reinhoudt, M. Crego-Calaman, What do we need for a superhydrophobic surface? A review on the recent progress in the preparation of superhydrophobic surfaces *Chem. Soc. Rev.*, **36**(8) 1350-1368 (2007)
[10]R. Blossey, Self-cleaning surfaces - virtual realities, *Nat. Mater.* **2** 301-306 (2003).
[11]H. Keskinen, J.M. Mäkelä, M. Aromaa, J. Keskinen, S. Areva, C.V. Teixeira, J.B. Rosenholm, V. Pore, M. Ritala, M. Leskelä, M. Raulio, M.S. Salkinoja-Salonen, E. Levänen, T. Mäntylä: Titania and Titania-Silver Nanoparticle Deposits Made by Liquid Flame Spray and their Functionality as Photocatalyst for Organic- and Biofilm Removal. *Catalysis Letters* **111**, 127-132 (2006).
[12]V. Pore, M. Ritala, M. Leskelä, S. Areva, M. Järn, J. Järnström: H2S Modified Atomic Layer Deposition Process for Photocatalytic TiO_2 Thin Films. *J. Mater. Chem.* **17** 1361-1371 (2007).
[13]M. Raulio, V. Pore, S. Areva, M. Ritala, M. Leskelä, M. Lindén, J.B. Rosenholm, K. Lounatmaa, M. Salkinoja-Salonen: Destruction of Deinococcus Geothermalis Biofilm by Photocatalytic ALD and Sol-Gel TiO_2 Surfaces. *J. Ind. Microbiol. Biotechnol.* **33**, 261-268 (2006).
[14]J. Määttä, M. Piispanen, H.-R., Kymäläinen, A. Uusi-Rauva, K-.R. Hurme, S. Areva, A.-M. Sjöberg, L. Hupa, Effects of UV-radiation on the cleanability of titanium dioxide-coated glazed ceramics, *J. Eur. Ceram. Soc.*, **27**(16), 4569-4574 (2007).
[15]J. Määttä, M. Piispanen, R. Kuisma, H.-R., Kymäläinen, A. Uusi-Rauva, K-.R. Hurme, S. Areva, A.-M. Sjöberg, L. Hupa, Effect on cleanbility of glazed surfaces, *J. Eur. Ceram. Soc.*, **27**(16), 4555-4560 (2007).
[16]R. Kuisma, L. Fröberg, H.-R., Kymäläinen, E. Pesonen-Leinonen, M. Piispanen, P. Melamies, M. Hautala, A.-M. Sjöberg, L. Hupa, Microstructure and cleanability of uncoated and fluoropolymer, zirconia and titania coated ceramic glazed surfaces, *J. Eur. Ceram. Soc.*, **27**(1), 101-108 (2006).
[17]M. Piispanen, L. Fröberg, T. Kronberg, S. Areva, L. Hupa, Corrosion of glazes coated with functional films in detergent solutions, Advances is Science and Technology, 45, 2006, 156-161
[18]M. Piispanen, J. Määttä, S. Areva, A.-M. Sjöberg, M. Hupa, L. Hupa, Chemical resistance and cleaning properties of coated glazed surfaces, *J. Eur. Ceram. Soc.*, **29** 1855-1860 (2009).
[19]M. Piispanen, J. Määttä, T. Kronberg, S. Areva, A.-M. Sjöberg, L. Hupa, Effect of mechanical and chemical wear on cleanability of functional and traditional glazed surfaces, Proceedings in 10th International Conference and Exhibition of the European Ceramic Society (June 17-21, 2007, Berlin), ISBN 3-87264-022-4
[20]M. Aromaa, H. Keskinen, J.M. Mäkelä, The effect of process parameters on the Liquid Flame Spray generated nanoparticles, *Biomol Eng*, **24**(5) 543-548 (2007).
[21]H. Keskinen, J.M. Mäkelä, M. Aromaa, J. Ristimäki, T. Kanerva, E. Levänen, T. Mäntylä, J. Keskinen, Effect of silver addition on the formation and deposition of titania nanoparticles produced by liquid flame spray, *Nanoparticle J. Res* **9** 569-588 (2007)
[22]Wenzel, R.N., Resistance of solid surfaces to wetting by water, *Ind.Eng.Chem.*, **28**, 988-994 (1936)
[23]Van Oss, C.J., Chaudhury, M.K., Goodm R.J., Interfacial Lifshitz-van der Waals and polar interactions in macroscopic systems, *Chem. Rev.*, **88**(6), 927-941 (1988)

[24] S. Takeda, K. Yamamoto, Y. Hayaska, K. Matsumoto, Surface OH group governing wettability of commercial glasses, *J. of Non-Cryst. Solids* **249**, 41-46 (1999)
[25] S. Takeda, M. Fukawa, Role of surface OH groups in surface chemical properties of metal oxide, *Mater. Sci. Eng. B* ***119***, 265-267 (2005)
[26] A. Kanta, R. Sedev, J. Ralston, Thermally- and Photoinduced Changes in Water Wettability of Low-Surface-Area Silica and Titania, *Langmuir*, **21**, 2400-2407 (2005)
[27] N. Stevens, C.I. Priest, R. Sedev, J. Ralston, Wettability of Photoresponsive Titanium Dioxide Surfaces, *Langmuir*, **19**, 3272-3275 (2003)

SKIN DAMAGE BY DIFFERENT SHAPES OF PHOTOCATALYST NANOPARTICLES

Eiji Watanabe, Kaoru Nishizawa, Takeshi Miki, and Hiroshi Taoda
National Institute of Advanced Industrial Science and Technology (AIST chubu)
Shimoshidami-Anagahora 2266-98, Moriyama-ku, Nagoya, 463-8560, JAPAN

ABSTRACT

We studied the influence of different titania photocatalyst nanoparticles shape on artificial skin similar to that of the human body under UV light irradiation conditions. The decomposition degree of the artificial skin was evaluated by monitoring the amount of carbon dioxide generated from the titania photocatalyst nanoparticle action. If titania nanoparticles are placed directly on the skin, the carbon dioxide generated from the skin was not detected in the titania nanoparticle coated with silica although it was detected in the sample that had no silica coating. The influence on the skin decreased because the titania nanoparticle coated partially in silica did not touch the skin directly.

INTRODUCTION

The titania photocatalyst has been used in various fields, and in particular is being used for environmental purification and protection of the living environment in recent years [1, 2, 3]. However, very little is still known about the environmental damage caused by the titania photocatalyst. This has been a cause of anxiety around world and has threatened to become a factor hindering the further widespread use of the titania photocatalyst [4]. Furthermore, new obligations for the restriction and safety evaluation of new and existing chemicals were recently imposed. A new index in addition to the past indicator of Organization for Economic Co-operation and Development (OECD) [5] has been proposed. The new chemical restriction method is called REACH (Registration, Evaluation and Authorization of Chemicals) in the European Union (EU) was enforced in spring 2007 [6]. The new REACH stipulations demand that a skin corrosion test and/or skin irritation test be implemented.

Against this background we attempted to construct a new evaluation method for damage to human skin. The decomposition degree of artificial skin and human skin was evaluated by monitoring the amount of carbon dioxide generated from them through titania photocatalyst nanoparticle activity [9]. The monitoring technique to evaluate the damage to the artificial skin by the titania photocatalyst nanoparticles aims to estimate the harmful strength of the photocatalyst to the environment and living things. This paper reports on the influence of different shapes of titania photocatalyst nanoparticles on artificial skin The artificial skin used in the experiment was "Bess chitin" which reacts similarly to human body under UV light irradiation conditions.

PROCEDURE

Testing equipment

The examination apparatus and analytical system were partially modified, with a test method based on a reactive container (photoreactor) and the measurement system as described in the ISO law, 22197-1 that was established in 2007 [8].

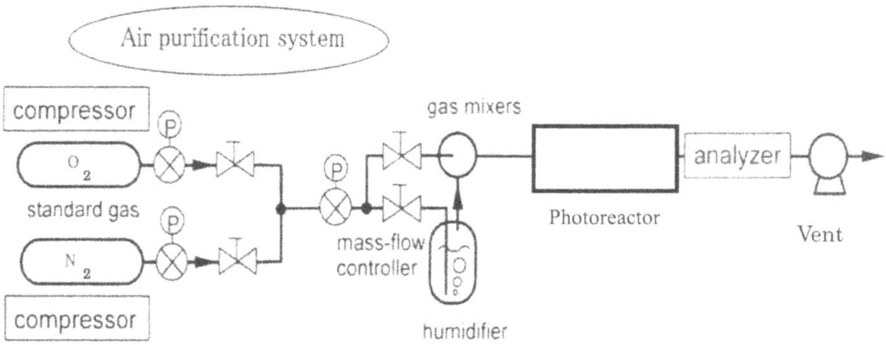

Figure 1. Testing equipment

The UV (ultraviolet ray) irradiation device and the photoreactor were placed in the shade box (at room temperature 25 degrees Celsius) as reported [9]. The UV rays were irradiated through a quartz glass plate above the photoreactor, and the quantity of light irradiation was adjusted to 1.0 $\mu W/cm^2$ on the sample surface. The damage on the skin tissue models sample by the quantity of UV ray has been understood to be able to disregard the influence from a preliminary experiment. The sample sheet (artificial skin) was placed on a disk of one centimeter in diameter.

There were connecting tubes of atmospheric gas at two places in the photoreactor. The composition of the input atmospheric gas was mixed steam so that humidity was kept at a constant 50% and oxygen was mixed with nitrogen at a ratio of 8:2 as a zero calibration gas based on the composition of the atmosphere. Flow rate was adjusted to 0.3L/min.

Basic operation of the examination method

The constant amount (0.01g) of the titania photocatalyst nanoparticles distributed to the culture solution by the supersonic wave was implemented by placing one drop with a syringe, uniformly sprinkled on the surface of the sample skin placed on the holder, and then packed closely under a glass window plate. The zero calibration gas flow was continued without irradiation until the baseline was stabilized. Ultraviolet rays were then irradiated from the upper photoreactor to the sample surface. Carbon dioxide released from the artificial skin was identified and measured by the gas analysis method (VA-3000type, HORIBA, Ltd.) at regular intervals during the irrigation.

RESULT AND DISCUSSION

Basic samples of the titania photocatalyst nanoparticle used predominantly in the Japanese market were used as the object photocatalyst nanoparticle (table 1. ST21 produced by Ishihara Sangyo Co. Ltd.). The test sample was a basic sample of titania nanoparticles partially covered with a silica net layer modified using a procedure developed by Taoda [9]. The shape of the sample was the bipolar system with which a part of titania surface exposed, as seen in figure 2.

Table 1. Titania photocatalyst nanoparticle basic sample

	Basic Sample
X-ray particle diameter (nm)	7
Specific surface area (m^2/g)	. 300

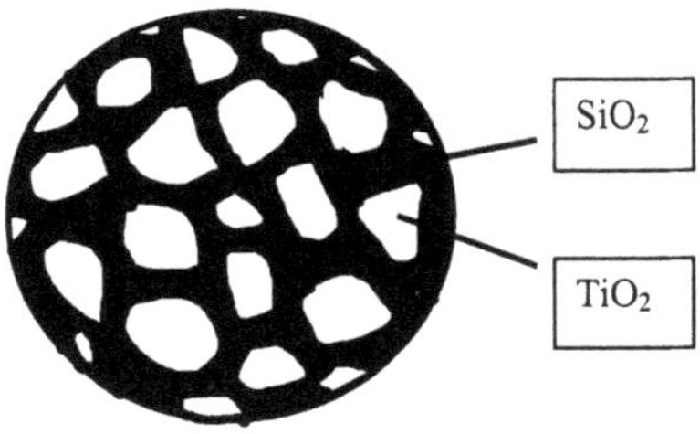

Figure 2. Surface of the test sample of titania nanoparticle partially covered with silica net layer

The artificial skin "Bess chitin" developed by Unitika Co. Ltd was used as an evaluation test sample [9]. Figure 3 shows the chemical constitution formula of Bess chitin. Figure 4 shows the photograph taken using a Color 3D Laser Scanning Microscope.

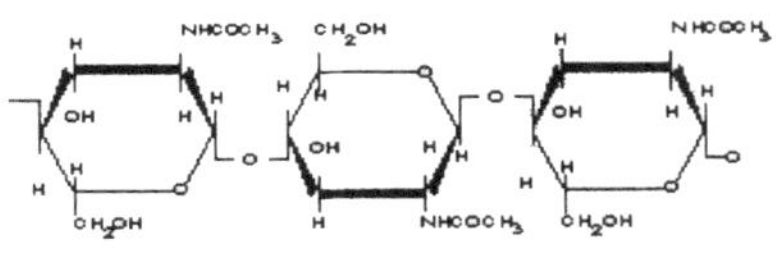

Figure 3. Chemical constitution formula of Bess chitin $(C_8H_{13}NO_5)_n$

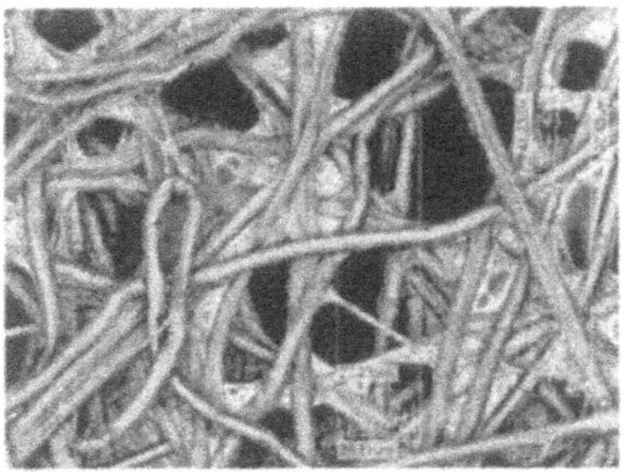

Figure 4. Photograph taken using Color 3D Laser Scanning Microscope

Changes in carbon dioxide concentration generated by different shapes of titania photocatalyst nanoparticles

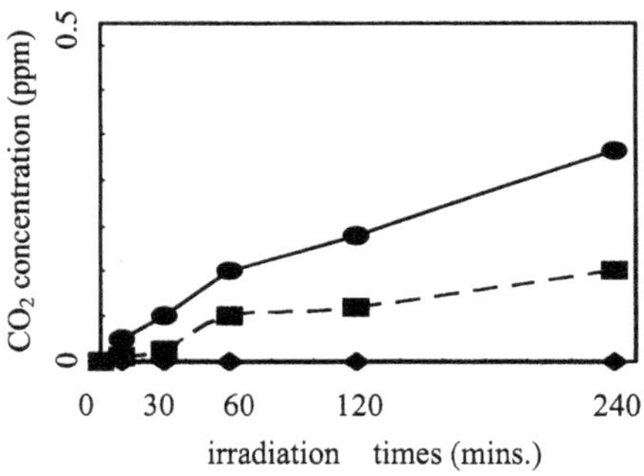

Figure 5 Changes of carbon dioxide concentration generated by different titania photocatalyst nanoparticle shapes
●: Basic sample ■: Test sample : Blank

Under UV irradiation conditions, the photocatalyst activity of the titania nanoparticle covered partially by a silica net layer (test ample) and the titania nanoparticles not covered (basic sample) were measured using a decomposition test for the organic matter (1 ppm of m-chlorobenzoic acid solution) with one-dropped of solution being placed on the titania powder placed on the glass plate.

Figure 5 shows that the amount of carbon dioxide generation with both samples tends to increase gradually. The generation level of the basic sample was inferior compared with the test sample. It seems that the difference in the amount of carbon dioxide generation between the two sample is related to the amount of the titania photocatalyst exposed to the surface. The carbon dioxide generated with the test sample decreases due to the fact that a part of the surface is covered with a silica net layer, reducing the contact of organic matter with the atmosphere.

Changes in carbon dioxide concentration generated from the artificial skin decomposed by titania photocatalyst

It is necessary to reveal what harmful influences the titania photocatalyst nanoparticle has on the environment and the human body skin. We used a test method using artificial skin, which is similar to human skin [9]. The damage caused by the titania photocatalyst particle to the artificial skin was examined by the concentration changes of carbon dioxide generated in gas atmosphere. The concentrations of carbon dioxide generated from the sample skins were monitored with time. It was not possible to detect any carbon dioxide in the experiment in which no sample was used, with the titania photocatalyst only being irradiated by UV light in the examination time. As a result, it was found that the change in concentration of the carbon dioxide generated in the gas atmosphere could be caused by the titania photocatalyst activity on the skin.

Changes in carbon dioxide concentration generated from the artificial skin decomposed by the different titania photocatalyst particle shapes

It was found that the influence of the titania photocatalyst to artificial skin could be evaluated by measuring the amount of carbon dioxide generated from the skin damaged by the titania photocatalyst activity [9]. The artificial skin materials were developed for regenerative medicine. The change in the carbon dioxide generated when the photocatalyst nanoparticle was spread on the epidermal sample under the UV lights irradiation could be measured.

Figure 6 shows that the photocatalyst activity on artificial skin of the titania nanoparticle that was partially covered by the silica layer (test sample) and the titania nanoparticles not covered (basic sample), when both samples are under UV light irradiation conditions. In the case of the basic sample, the concentration of carbon dioxide increased rapidly when the UV light was switched on after which it maintained a contact concentration for the duration of the irradiation. It decreased rapidly when the light was switched off after which it could hardly be detected.

It is thought that the carbon dioxide separated and released in the gas atmosphere as the skin in the basic sample decomposed by photocatalyst activity under UV irradiation conditions. The experiment showed that carbon dioxide was generated from the artificial skin in the basic sample through photocatalytic activity.

On the other hand, no carbon dioxide was detected in case of test sample. This suggests that there was no influence on the skin because the titania photocatalyst nanoparticle was not in direct contact with the skin in this sample.

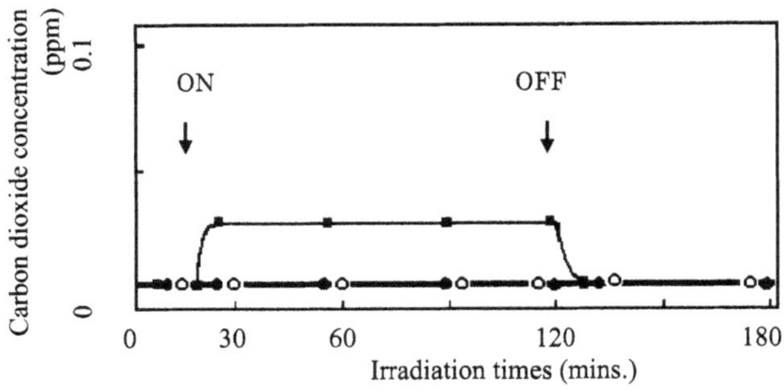

Figure 6. Changes in carbon dioxide concentration generated from the sample decomposed by titania photocatalyst

○: Blank ■: Basic sample ●: Test sample

ON: UV light on OFF: UV light off

Because the basic sample was in direct contact with the skin, the titania photocatalyst activity had an influence on the skin. The shape of test sample was a bipolar system for which a part of titania nanoparticle surface was covered in a silica net layer. Neither an outside structure (skin etc.) nor titania particles come in contact directly for this reason. As a result, the photocatalyst activity of titania was interrupted by the silica coating layer and would not reach the skin. It seems that the influence on the skin disappeared for such a bipolar system though the photocatalyst performance decreased a little.

Whether the artificial skin had actually received damage by the photocatalyst action was examined using instrumental analysis (e.g. FE-SEM observation and so on). It was not founded the change on the surface shape of the artificial skin during the UV light irradiation of 120 minutes when observing it by instrumental analysis. It was thought that change of the artificial skin surface decomposed by the photocatalyst activity would not largely take place in a short time frame. However, there is a possibility that a change to the surface shape of the artificial skin takes place when long term irradiation is continued.

Figure 7 shows that changes in carbon dioxide concentration were not detected for all samples under visible light irradiation conditions. Although all samples were in direct contact with the skin, the titania photocatalyst activity could not influence the skin. It seemed that the influence on the skin by the photocatalyst activity could not occur because the basic titania photocatalyst did not react under visible conditions.

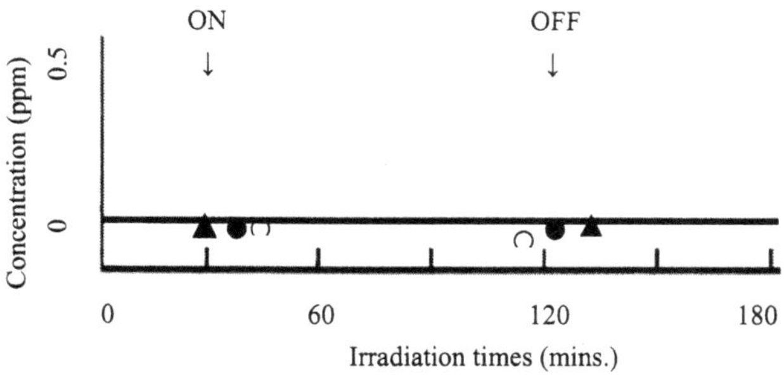

Figure 7. Changes in carbon dioxide concentration generated from the sample decomposed by titania photocatalyst under visible light
○: Blank ▲: Basic sample ●: Test sample
ON: Visible light on OFF: Visible light off

CONCLUSION

The influence of titania photocatalyst nanoparticles on the skin is to be doubted. We attempted to examine the influence of titania photocatalyst nanoparticles with different shapes on artificial skin similar to human skin under the UV light irradiation conditions. Carbon dioxide was released in the gas atmosphere from the artificial skin decomposed by the original titania photocatalyst action. On the other hand, carbon dioxide generated from the skin was not detected in the titania photocatalyst nanoparticle coated partially by silica layer. The photocatalyst action of titania did not reach and decompose the artificial skin because the titania particle and the artificial skin did not come in direct contact. In case of the combination of materials, for example silica and silica without the photocatalyst action, and partially covered titania particle surface, the titania photocatalyst action did not influence the artificial skin.

Under visible conditions, the titania photocatalyst could not influence the skin because the photocatalyst activity could not generated under visible conditions.

REFERENCES

[1]F. Mashio and S. Kato, Study on oxidation reaction used titania photocatalyst, *J. Ind. Chem. Soc.*, **67,** 1136-1141(1959)

[2]K. Honda and A. Fujishima, Electrochemical photolysis of water at a semiconductor electrode, *nature,* **238**, 37-38(1972)

[3]E. Watanabe, H. Taoda, K. Kato, T. Nonami, K. Iseda, and M. Fukaya, Water purification by titanium dioxide thin film photocatalyst, *Reports of NIRIN*, **46**, 227-238(1997)

[4]L. Thrall, Study links TiO_2 nanoparticles with potential for brain-cell damage, *Environmental Sci. & Tech. online News* (2006)
[5]OECD, Guidelines for the Testing of Chemicals, (1992)
[6]REACH report, Information on http://ec.europa.eu/environment/chemicals/reach/reach_intro.htm, (2008)
[7]H. Taoda, Environmental clean-up technology used ceramics photocatalyst, *J. Solar Energy Soc.*, **26**, 13-15(2000)
[8]International Organization for Standardization, 22197-1, Test method for air purification performance of semiconducting photocatalytic materials, Part 1: Removal of nitric oxide, (2007)
[9]E. Watanabe, M. Fukaya, K. Nishizawa, T. Miki, and H. Taoda, Test method for skin damage of titania photocatalyst nanoparticles, *Materials Sci. Forum*, **569**, 9-12(2008)

CHARACTERIZATION OF CRYSTALLIZED TiO_2 FILM BY POST PLASMA TREATMENT

Toshiya Watanabe[a], Yukinobu Yokota[a], Naoya Yoshida[a], Yuko Shibayama[a] and Hisashi Ohsaki[b]
[a] The University of Tokyo,4-6-1,Komaba,Meguro-ku,Tokyo 153-8904 Japan
[b] 1-1-1 Umezono Tsukuba, Ibaraki 305-8568 Japan

ABSTRACT

The effect of post-plasma treatment on the surface roughness and wettability of a TiO_2 film deposited by magnetron sputtering was studied. The crystallization of anatase by the plasma treatment was optimized by the selection of an amorphous thin film that was deposited at 20% oxygen partial pressure. The anatase film crystallized by the plasma treatment had a flatter surface with a lower average roughness than a heat-treated sample.

The flat surface of the thin film exhibited improved hydrophobicity. Although the highest value of the water contact angle was around 70°, the film showed excellent water removability. The sliding angle of the water droplet was only 20°. Microstructure of the film with higher water removability was examined by TEM observation and it was clarified that an amorphous top layer was formed on the crystallized layers.

It is shown that a TiO_2 photocatalyst surface composed by amorphous top layer with nanometer scale flatness exhibits superior dynamic hydrophobicity.

INTRODUCTION

Titanium dioxide (TiO_2) has been used in various products because of its photocatalytic functions. Its self-cleaning function based on not only photocatalytic oxidation but also photo-induced hydrophilic conversion has attracted much attention. There are two types of photochemical reactions that occur on the surface of TiO_2 when it is irradiated with ultraviolet light. The first type is the photo-induced redox reaction of adsorbed substances, and the second type is the photo-induced hydrophilic conversion of TiO_2 [1]. It was found that a combination of these two reactions is very effective for self-cleaning purposes. For instance, due to these reactions, oily stains on a highly hydrophilic building surface could be removed easily by the occasional rainfall. Because of these novel findings, the application of TiO_2, particularly in the field of building materials, has been accelerated [2].

However, hydrophilic surface has some negative effects on the cleaning function. It is known that a hydrophilic surface is thermodynamically unstable; therefore, it adsorbs organic gases. Furthermore, stains easily form on its surface, and although it can be easily cleaned by water, stains tend to remain on it. In regard to this point, photo-induced hydrophilic surface of TiO_2 is not an ideal one, furthermore should be improved especially when it is used where enough rain cannot be expected. Otherwise, such thermodynamically unstable surface results in more contamination than the usual exterior material does.

To find a solution to this problem, we tried to develop hydrophobic TiO_2 surface with low surface energy that exhibit photocatalytic activity. As mentioned already, TiO_2 usually exhibits a rather hydrophilic state with a contact angle of around 30°. In order to obtain higher hydrophobicity, it is necessary to modify its surface.

According to Young's equations, surfaces with lower energy exhibit higher water repellency. However, in the case of an extremely hydrophobic surface such as the surface of a lotus leaf, the contact angle cannot be predicted by only Young's equations; rather by using a different law.

There have been many reports on superhydrophobic surfaces that are based on the Cassie-Baxter law [3]. However, only a film covered with hydrophobic polymer such as fruorocarbon shows a high water contact angle, more than 120°, when its surface is roughened significantly. Roughened surfaces with polymer typically show less mechanical durability than flat surfaces of inorganic material. Because of such problems, little commercialization of superhydrophobic film has been conducted.

In this study, a quite different approach for obtaining a hydrophobic photocatalytic surface without roughened surface or polymer is performed. The key concept is based on a classical model known as the Wenzel law. According to the modified Young's equation derived by Wenzel [4], the roughness factor enhances both hydrophilicity and hydrophobicity of a surface. According to the Wenzel law, when the intrinsic water contact angle on a flat surface is larger than 90°, the hydrophobicity of the surface increases. However, when the intrinsic water contact angle on the flat surface is smaller than 90°, the hydrophilicity of the surface increases. It means that flatter surface of inorganic material is expected to show more hydrophobicity.

In this study, we attempted to develop TiO_2 surface with high level of flatness to increase hysrophobicity. To the best of our knowledge, although there are papers describing the effect of surface roughness on the water contact angle [5], there are no reports that describe the effects of nanometer-scale roughness. In this study, we have attempted to improve the hydrophobicity of a TiO_2 film by optimizing the nanometer-scale roughness through post-plasma treatment; this film was prepared by the magnetron sputtering process.

Generally, the physical deposition process produces a denser film than a wet process such as the sol-gel process. In order to obtain a TiO_2 surface that is flat on a nanometer scale, the sputtering process is useful because it is considered to be superior to the wet process. However, sputtering at room temperature cannot produce crystallized TiO_2. To obtain a high photocatalytic TiO_2 thin film, heat treatment at an elevated temperature is necessary. However, this process not only increases processing time but also causes grain growth which leads to increase roughness [6].

The post-plasma treatment of an amorphous oxide thin film for low temperature crystallization has been studied by our group. Through this process, a crystallization technology using RF plasma was developed. Various amorphous films were successfully crystallized and densified by treating them in less than 5 min with a slight increase in the film temperature [7].

In this study, post-plasma treatment was used to produce a photocatalytic TiO_2 thin film having a nano-scale flat surface and high hydrophobicity.

EXPERIMENTAL

An amorphous TiO_2 thin film was prepared by the sputtering method. TiO_2 was deposited by using a Ti target and a sputtering gas of 3% O_2 diluted with Ar and a DC power supply (Advanced Energy Inc., MDX). The substrate temperature did not exceed c.a. 30 °C during the deposition process, and amorphous TiO_2 films were obtained. The samples were precoated with a SiO_2 alkali barrier, which was deposited from a quartz target in an Ar-based sputtering gas containing 5% O_2, by using a solid-state 13.56-MHz RF power generator (ENI Technology, Inc., OEM-12A) and an auto-impedance matcher (ENI Technology, Inc.).

The deposited films were treated with capacitively coupled 13.56-MHz RF plasma in a barrel-type quartz or borosilicate chamber having a pair of electrodes, each covering half of the cylinder's wall. RF power was supplied by a solid-state 13.56-MHz RF power generator (Huettinger Elektronik GmbH + Co. KG, QINTO 3013) through an auto impedance matcher (RF VII, Inc.). The film temperature during the plasma treatment was measured by using a radiation thermometer, and its reproducibility was about ±4 °C. The plasma crystallization process was carried out at an oxygen pressure of 100% (225 sccm, 330 Pa) and an RF power of 500 W. During plasma treatment, Cu electrode is attached to the backside of substrate and electrically floated from the electric ground. It intensifies RF electrical field and enhances the plasma crystallization performance. Some of the samples are conducted by plasma treatment without Cu electrode to compare the properties. For comparative study, some of the sample deposited by sputtering process were treated at elevated temperature at 500 ℃ for 30 min.

The crystalline structure of the films was evaluated by the asymmetric XRD method employing the Seemann-Bohlin arrangement with a fixed X-ray incident angle of 0.8°. The XRD machine was a RINT-2100 (Rigaku Co.) with CuKα operated at 40 kV and 30 mA. AFM measurements were also conducted.

The photocatalysis was evaluated by carrying out the degradation of gaseous isopropanol (IPA) under 1.0-mW/cm^2 UV-light irradiation (Toshiba Lighting & Technology Co., black light bulb (BLB)). The film sample was placed in a 1000-ml Pyrex vessel, and the initial IPA concentration was set at about 1000 ppm. The UV intensity was measured using a UV power meter (TOPCON Co., UVR-2). The concentrations of IPA and acetone, which was formed as an intermediate in the conversion from IPA to CO_2 and H_2O, were measured using a gas chromatograph (Shimadzu Co., GC-8A) that was equipped with a 5-m PEG 1000 column and a flame ionization detector with N_2 as the carrier gas. In addition, the CO_2 concentration was monitored using GC-8A with a 2-m active-carbon column and a methanizer and a flame ionization detector using N_2 as the carrier gas. Surface wettability was determined by measuring the water contact angle by the sessile drop method using a contact angle meter (Kyowa Interface Science Co., Ltd., CA-X).

TEM observation was carried out by using H9000NAR (acceleration voltage is 300kV).

RESULTS

Figure 1 shows the relationship between the oxygen partial pressure of sputtering and the height of the anatase peak of the samples after plasma treatment with Cu electrode. The thin film deposited at an oxygen partial pressure of less than 20% was not crystallized by the plasma treatment, and Ti metal was detected by X-ray analysis. At an oxygen partial pressure of more than 20%, a broad rutile peak was detected, but no anatase peak was observed. An increase in the oxygen partial pressure led to a decrease in the amount of the anatase phase and no crystallization occurs after plasma treatment. From these results, the optimum oxygen partial pressure required for producing an anatase-type TiO_2 film was estimated to be around 20%. On the other hands, plasma treated samples without Cu electrode does not show such crystallization. Only broad anatase peak was detected at an oxygen partial pressure of 20%.

Photocatalytic activity of TiO_2 film optimized oxygen partial pressure (20%) by IPA decomposition was examined. It was clarified that the anatase TiO_2 film which was crystallized by the plasma treatment shows high photocatalytic activity. It is almost same with that of heat treated one (Figure 2).

Figure 3 shows the average roughness of the fabricated thin film measured by AFM (scanned area = 0.5 μm^2). A clear dependency of the oxygen partial pressure on roughness can be seen. This is attributable to the dependency of the oxygen partial pressure on the crystallization of anatase, as shown in Fig. 1. At any partial pressure, the plasma-treated samples showed lower roughness than the heat-treated samples.

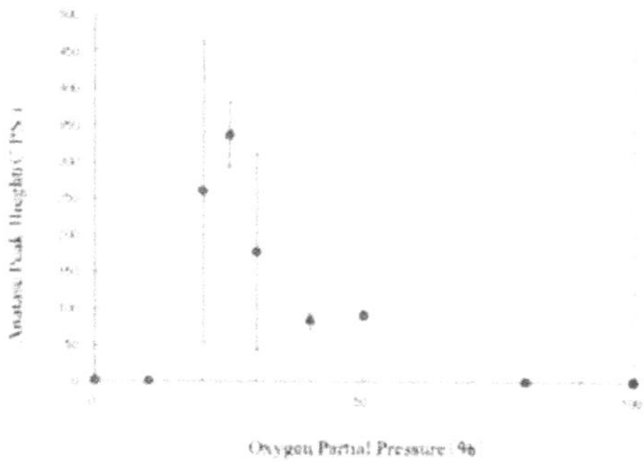

Figure 1. Effect of oxygen partial pressure of sputtering on anatase phase after plasma treatment. X-ray diffraction measurements were conducted on several samples. Error bars show standard deviations in measurements.

The effect of the duration of plasma treatment on the surface roughness of the TiO_2 film was also studied. Longer treatment durations resulted in a slight increase in roughness. The deviation in the roughness of the sample treated by plasma for longer durations tended to be larger. In our previous study, crystallization occurred within several minutes under the optimal plasma treatment. Therefore, the optimum duration of plasma treatment for obtaining a flat surface was considered to be around several minutes.

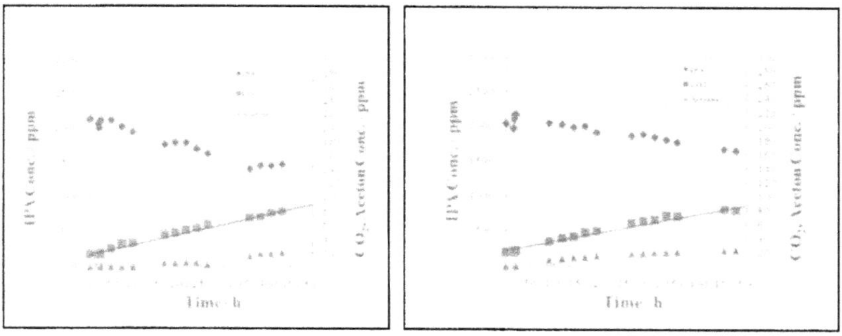

(a) (b)

Figure 2. Evaluation of photocatalytic activity of TiO_2 film by IPA decomposition tests

(a) Sample heat treated at 500 ℃ for 30 min.

(b) Sample treated by plasma at room temperature

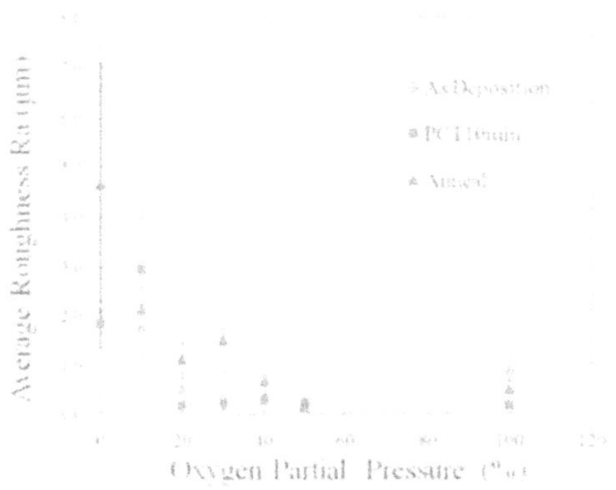

Figure 3. Dependence of oxygen partial pressure of sputtering on average surface roughness of TiO_2 film in the processes of deposition, plasma treatment (PCT) and heat treatment (Anneal). Plasma treatment time was 10 min and heat treatment time was 30 min.

Figure 4 shows AFM images of the TiO_2 thin film prepared in the optimized condition. It is clear that the surface of the thin film has low roughness and few projections. Figure 5 shows SEM images of these samples. 10-nm sized grains are clearly observed on the heat-treated surface. The observed microstructure and grain size are almost identical with those found in previous studies [8]. However, clear grains and grain boundaries are not observed in SEM images of plasma treated surfaces. Although the microstructures of the TiO_2 film prepared by the magnetron sputtering process were studied by many researchers [9], such a highly flat surface with fine grains has not yet been studied. The characteristics of the flat and fine microstructure of the TiO_2 film prepared by the post-plasma process without an apparent temperature rise are considered unique.

Because of the heat flow, the crystallization and densification of the heat-treated TiO_2 film should commence from the surface and the sides of the substrate. Nucleation also occurs according to the heat flow. However, in the plasma-crystallized TiO_2 film, crystallization commences throughout the film because RF electromagnetic waves supply the energy for crystallization [10]. The crystal nucleation sites are more abundant in the plasma treatment process than in the heat treatment process. The smaller grain size of the plasma-treated samples is attributable to the increase in the number of nucleation sites in the film during crystallization.

It was also found that Cu electrode attachment to intensify RF electrode does not affect on the microstructure and average roughness.

(a) (b)

Figure 4: AFM images of prepared TiO_2 surfaces

(a) Heat treated at 500 ℃

(b) Treated by plasma

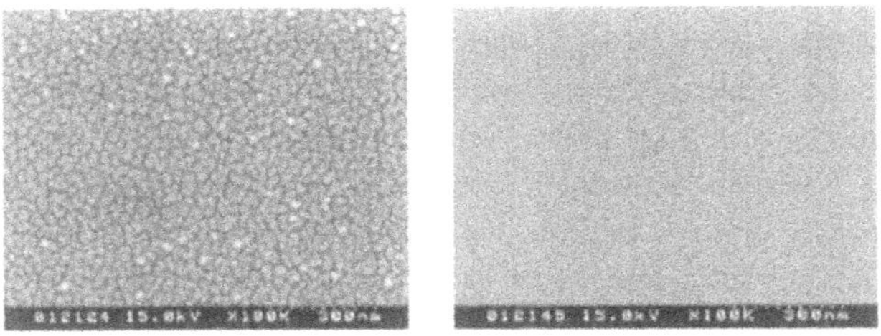

(a) (b)

Figure 5. SEM images of prepared TiO_2 surface

(a) Heat treated at 500 ℃ (b) Treated by plasma

Figure 6 shows the water contact angle of the prepared samples. The square plots indicate those of the plasma-treated samples, and the circular plots indicate those of the heat-treated samples. The figure shows the effect of a flatter surface on the improvement of the water contact angle. The highest water contact angle is more than 60° for a sample whose average roughness is less than 1 nm.

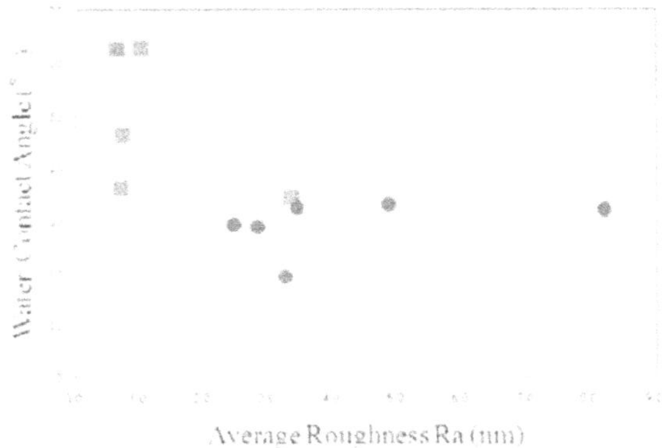

Figure 6. Relationship between the average surface roughness and the water contact angle. The red circular plots indicate those observed for heat-treated samples and the blue square plots indicate those observed for plasma treated samples.

This suggests that the surface wettability depends on surface roughness even on the nanometer scale. A roughness of less than 1 nm improves hydrophobicity significantly. Moreover, the sample whose water contact angle was the highest showed superior water removability. The water droplet on its surface slipped at a slide angle of only 20°. Figure 7 shows the shape of the water droplet sliding off the optimized TiO_2 surface. It should be noted that this photograph shows that contact angle hysteresis (difference between the receding angle and the advancing angle) is very small; this indicates a higher dynamic hydrophobicity. It is also noted that although the water contact angle of the sample without Cu electrode is almost same with them with Cu electrode, no such higher water removability was observed. Thus, it was clarified that the plasma treatment with Cu electrode gives samples with nanometer scale flatness and they show not only improved static hydrophobicity but also dynamic hydrophobicity.

Although it is apparent that higher contact angle of the TiO_2 is because of the nanometer scale surface flatness, the reason why the fabricated TiO_2 shows higher water removability cannot be attributable only to surface flatness. Actually, the sample treated by plasma without Cu electrode does not show water removability even though its surface is as flat as other samples treated with Cu electrode. To clarify the mechanism of improved water removability, we have conducted TEM observation on these samples.

Figure 7. Shape of a water droplet sliding on the TiO_2 film treated by plasma. The tilting angle is 30°.

Figure 8 (a) shows TEM image of heat treated sample. It is clear that the film is composed by 50 – 100 nm crystallized grains. Figure 8 (b) shows TEM image of a plasma treated sample without Cu electrode. There are no such large grains as like in Figure 8 (a). However thin and small crystallized portions are observed at near the surface. There is columnar structure which is located at a distance of 10 to 20 nanometers. Inner layer of this film is almost amorphous. Figure 8 (c) shows TEM image of the TiO_2 film treated by plasma with Cu electrode. As mentioned earlier, Cu electrode has an effect to intensify electrical field of application. Inner layer of the thin film in Figure 8 (c) is composed with well crystallized grains. The size of grains is similar to that of heat treated one. However, amorphous thin layer is observed at near the surface. The thickness of the layer is approximately 7 to 9 nanometers. The columnar structure which is located at a distance of 10 to 20 nanometers is also observed in Figure 8 (c).

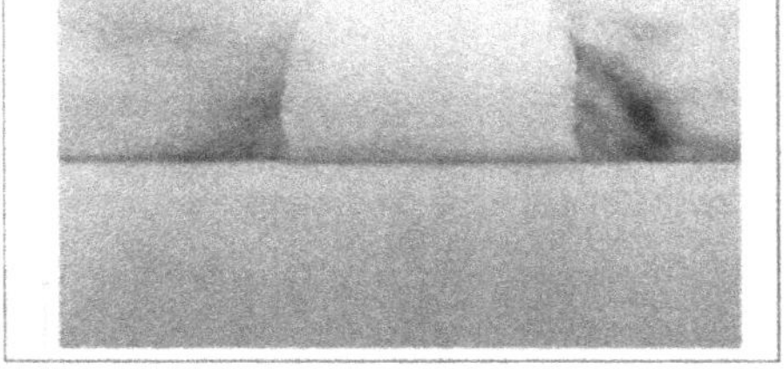

(a)

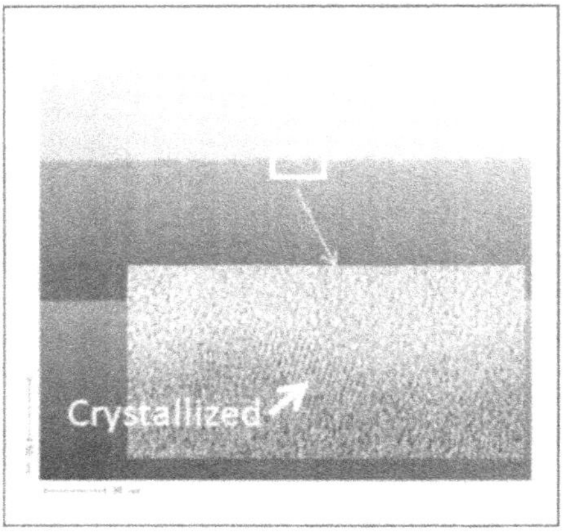

(b)

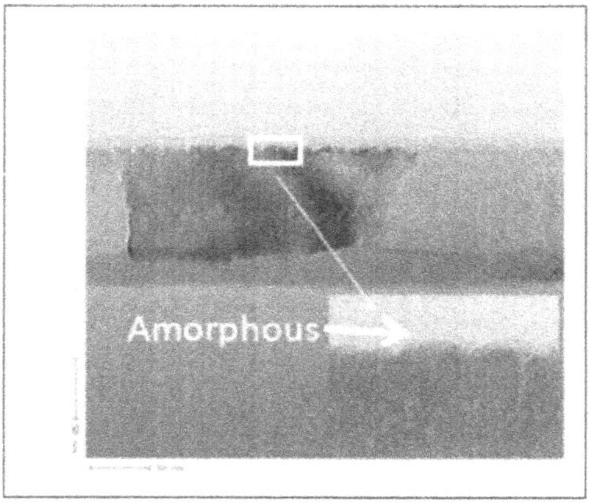

(c)

Figure 8. TEM photographs (a) Heat treated sample, (b) Plasma treated sample without Cu electrode, (c) Plasma treated sample with Cu electrode

Both of plasma treated samples shown in Figure 8 (b) and Figure 8 (c) exhibit similar water contact angles around 60 degree. However water removability is quite different between them. Although, water removability of the sample represented by Figure 8 (c) is around 20 degree, water droplet on the sample represented by Figure 8 (b) cannot be removed even when tilting angle is 90 degree. The difference can be attributable to the degree of crystallization of the surface layer.

DISCUSSIONS

Our results indicate that nanometer-scale roughness clearly has an effect on the water contact angle. Although this effect is identical to Wenzel law [4], the scale of the roughness is much smaller than that assumed in the Wenzel law. The Wenzel equation is fundamentally based on a thermodynamic model and the assumption that the average surface energy is identical to that enhanced by surface area. A surface with a roughness on a micrometer scale can be treated as a macroscopically uniform surface when it interacts with water droplets. However, surfaces that have a roughness in the nanometer scale may interact with water molecules in a different way. These surfaces have a typical roughness that is comparable to the size of water molecules and may be smaller than water clusters. In this case, a thermodynamically uniform interaction may not be expected.

Furthermore, it was clarified that the surface of TiO_2 sample which shows higher water removability was not crystallized but amorphous. Amorphous layer has no grains and grain boundaries which results in more homogeneous surface. In a sense, it is more heterogeneous not only physically but also chemically because it has no crystal structure. The chemical homogeneity caused by amorphous surface may cause water removability. To increase water removability, activation energy of the de-wetting process of water droplet should be decreased. Roughness is one of parameters to increase activation energy of de-wetting process. Furthermore, it is considered that the activation energy depends on chemical and physical homogeneity. Therefore, amorphous surface layer can contribute to decrease activation energy of de-wetting process.

Although further study on the above mechanism will be necessary, the present study revealed that the nanometer scale flat surface of TiO_2 with amorphous top layer exhibits enhanced hydrophobicity. We observed not only static hydrophobicity but also water removability. The fact that the wettability of a TiO_2 surface can be controlled to produce a higher dynamic hydrophobic state is useful for expanding the applications of TiO_2 photocatalytic thin films.

CONCLUSION

In order to improve the hydrophobicity of TiO_2 thin-film surfaces, we studied the microstructure and water wettability of thin films prepared by post-plasma treatment of amorphous TiO_2 deposited by the magnetron sputtering process. It was revealed that plasma treated samples exhibit lower roughness as compared to heat-treated samples. The enhancement is attributed to uniform crystallization occurring during the post-plasma treatment.

In this study, it was revealed that an optimized plasma treated sample exhibits a larger water contact angle, greater than 60°, and a lower slide angle. We found that a TiO_2 photocatalyst surface with nanometer scale flatness exhibits a superior hydrophobicity.

ACKNOWLEDGMENT

This work was performed under the management of Project to Create Photocatalyst Industry for Recycling-oriented Society supported by New Energy and Industrial Technology Development Organization (NEDO).

REFERENCES

[1] R. Wang, K. Hashimoto, A. Fujishima, M. Chikuni, E. Kojima, A. Kitamura, M. Shimohigoshi, and T. Watanabe, Nature 388 (1997)431.
[2] A. Fujishima, X. Zhangb, and D. A. Tryk; Surface Science Reports 63 (2008) 515–582
[3] A. B. D. Cassie,Discuss. Faraday Soc. 3(1948) 11.
[4] R. N. Wenzel, J. Phys. Colloid Chem.53 (1949) 1466.
[5] H. Irie, K. Tsujia, and K. Hashimoto, Phys. Chem. Chem. Phys. 10(2008) 3072.
[6] N. Kanai, T. Nuida, K. Ueta, K. Hashimoto, T. Watanabe and H. Ohsaki, Vacuum 74(2004) 723.
[7] H.Ohsaki, M.Suzuki, Y.Shibayama, A.Kinbara and T.Watanabe, J. Vacuum Sci. Technol. A 25(2007) 1052.
[8] Q. Ye, P. Y. Liu, Z. F. Tang, and L. Zhai,Vacuum 81 (2007) 627
[9] S.-H. Kim, Y.-L. Choi, Y.-S. Song, D. Y. Lee, and S.-J. Lee, Materials Letters 57(2002) 343
[10] H.Ohsaki, Y.Shibayama, M.Suzuki, A.Kinbara, H.Yashiro and T.Watanabe, "Room Temperature Crystallization by RF Plasma", Thin Solid Films 516, 4490-4494(2008).

Direct Thermal to Electrical Energy Conversion Materials

THERMOELECTRIC PROPERTIES OF OXYGEN DEFICIENT $La_{2-x}Sr_xCuO_{4-\delta}$ CERAMICS

Julio E. Rodríguez

Department of Physics
Universidad Nacional de Colombia
Thermoelectric Materials group
Apartado Aereo 85814
Bogotá, Colombia

ABSTRACT

Oxide ceramics with a nominal composition $La_{2-x}Sr_xCuO_{4-\delta}$ ($0<x\leq0.2$) (LSCuO) were grown by using the solid-state reaction method. The transport properties of the obtained samples were modified by submitting the samples to a post-growth high temperature thermal process. The thermoelectric properties were studied from Seebeck coefficient S(T), electrical resistivity ρ(T) and thermal conductivity κ(T) measurements. These transport properties were studied in the temperature range between 100 and 290K. The magnitude of S(T), ρ(T) and κ(T) decreases with the Sr doping level reaching maximum values close to 300 μV/K, 4 mΩ-cm and 4.7 W/K-m, respectively. From S(T), ρ(T) and κ(T) data it was possible to calculate the thermoelectric power factor (PF) and the dimensionless figure of merit (ZT), which reach maximum values close to 80 $\mu W/K^2$-cm and 0.4, respectively. The observed values in these performance parameters become these oxide ceramics promissory thermoelectric compounds for low temperature thermoelectric applications.

INTRODUCTION

The discovery of large positive Seebeck coefficient in the metallic oxide Na_xCoO_2 has shown the great potentiality of oxides as thermoelectric materials for a wide range of temperature applications. Since then, numerous studies have been devoted to the research of thermoelectric properties of different families of oxides. Taking in to account that for thermogeneration applications, P-type and N-type materials with similar properties are needed, the search for large Seebeck coefficient, positive or negative in oxide ceramics is very active [1-3]. In this sense, compounds as: $LaCoO_3$, $La_{1-x}Sr_xCoO_3$, $Bi_2Ca_2Co_2O_x$, etc., are promissory candidates to become thermoelectric materials, because of their transport properties and their physical-chemical stability [4-7].

$La_{2-x}Sr_xCuO_{4-\delta}$ compounds are members of perovskites-family, they adopt a tetragonal-symmetry and K_2NiF_4-type structure. This structure consists of layers of corners sharing BO_6 octahedra separated by AO rock salt layer. The majority of the transport phenomena in these compounds take place throughout the CuO_2 planes, which causes a marked asymmetry in their transport properties and a metallic or semiconducting behavior, which depend on the Sr content and critically on the oxygen stoichiometry [8].

Taking into account the versatility of their transport properties, these LSCuO ceramics can be considered as candidates for thermoelectric applications. In this sense, here, a study of thermoelectric properties of oxygen deficient LSCuO compounds is presented as a function of temperature and Sr content.

EXPERIMENT

Samples with nominal composition $La_{2-x}Sr_xCuO_{4-\delta}$ ($0<x\leq0.2$) were prepared by using ceramic routes from high purity powders of La_2O_3, SrO and CuO (Merck 99.99% pure). The oxygen content

present in the obtained samples was modified by submitting them to a thermal process at 900^{0}C during 50 hours in an inert atmosphere.

Seebeck coefficient data were obtained by differential technique, with an accuracy about 0.5 μV/K. The longitudinal heat flow method was used in the thermal conductivity measurements. To avoid heat loss via residual gases, κ(T) measurements were carried out under vacuum atmosphere (10^{-3} mb). Additionally, the measured thermal conductivity was corrected against radiation losses based on $\sim T^3$ temperature dependence [9,10]. The accuracy of thermal conductivity data was about 0.2 W/K-m.

On the other hand, the electrical resistivity was measured by the standard DC four probe method. These transport properties measurements were carried out in the temperature ran ge between 100 and 290K. In addition, the structural and morphological properties of the samples were studied by powder x-ray diffraction analysis (XRD) and Scanning Electron Microscopy (SEM), respectively.

DISCUSSION

The x-ray diffraction analysis shows a compound with tetragonal symmetry having a space group I4/mmm. No evidence of impurities or secondary phases was found in the spectra (see figure 1). The obtained lattice parameters do not change significantly with the Sr level, as can be seen in table 1.

Table 1. Lattice parameters of strontium doped LSCuO compounds obtained from powder x-ray diffraction analysis.

Sample	a=b (Å)	c (Å)
Sr=0.05	3.790(1)	13.299(6)
Sr=0.10	3.783(1)	13.278(5)
Sr=0.15	3.790(1)	13.314(5)
Sr=0.20	3.789(1)	13.302(6)

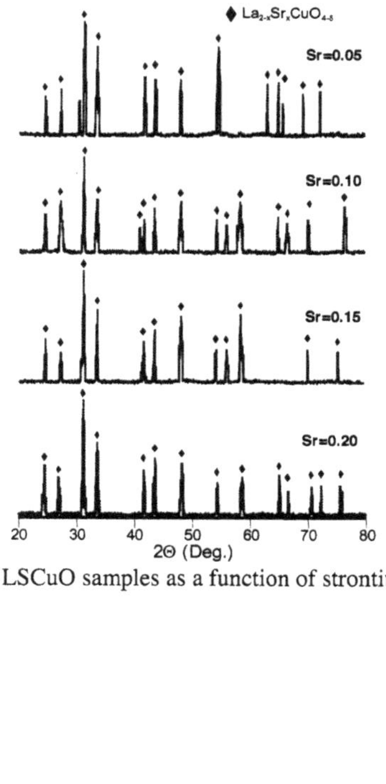

Figure 1. XRD patterns of LSCuO samples as a function of strontium content.

Figure 2. SEM micrographs of LSCuO compounds grown by solid-state reaction method.

Figure 2 shows the morphology of LSCuO compounds obtained by scanning electron microscopy. The samples show a well established grain structure, low-doped samples exhibit the grain structure and the porosity typical of these ceramic compounds, with increasing Sr content the porosity decreases and the grain size decreases too, which have effects on the thermal transport across the samples.

The temperature behavior of Seebeck coefficient is shown in the figure 3a, S(T) is positive over the measured temperature range, suggesting a hole-type conduction. Its magnitude decreases monotonically with the manganese level from 300 μV/K to 50 μV/K.

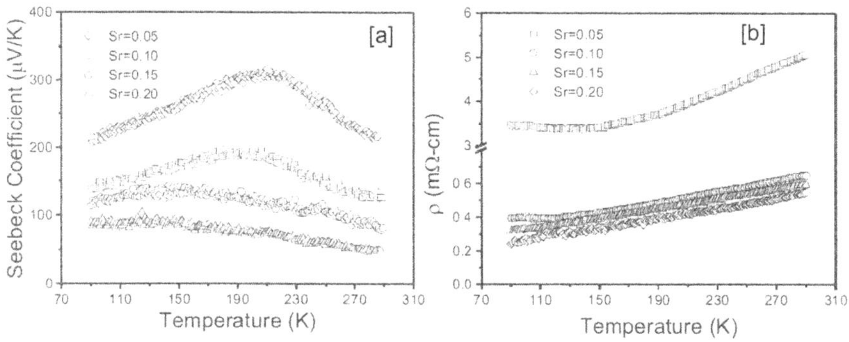

Figure 3. Temperature behavior of Seebeck coefficient [a] and electrical resistivity [b] for LSCuO ceramics as a function of temperature and the strontium content.

Across the measured temperature range, S(T) shows two different trends. Between 200 and 290K S(T) decreases weakly with temperature, this behavior can be described by Heikes model for strong correlated systems. According to this model S(T) is given by [11-13]

$$S - S_0 = \frac{k_B}{e}\ln\left[\frac{1-p}{p}\right] \approx \frac{E_g}{2eT} \quad (1)$$

where k_B, p (filling factor) and E_g are the Boltzmann constant, the electronic (hole) concentration and the energy gap, respectively.

By fitting S(T) experimental data to the Heikes model (plotting S(T) vs. 1/T) the values for E_g were obtained which are shown in table 3. The energy gap (E_g) decreases with the Sr level as a consequence of the increase of charge carriers density generated by the addition of strontium.

Table 3. Behavior of E_g parameter obtained from fitting Seebeck coefficient experimental data to $S - S_0 = \frac{k_B}{e}\ln\left[\frac{1-p}{p}\right] \approx \frac{E_g}{2eT}$ model.

Sample	Sr=0.05	Sr=0.10	Sr=0.15	Sr=0.20
E_g (meV)	112.6(2)	60.3(1)	43.8(2)	25.5(2)

In the temperature region between 100 and 200 K, the Seebeck coefficient is temperature dependent; this behavior can be ascribed to variable range hopping (VRH), in which S(T) is expressed as [12,13]

$$S(T) = AT^n \tag{2}$$

where A and n are a fitting parameter and the VRH exponent, respectively.

The obtained A and n parameters are presented in table 4, the increase of A parameter with the Sr doping make evident that the localized states are increasing with the Sr level. On the other hand, n parameter took values from 0.5 to 0.3, making evident the change in the dimensionality of the samples from 3D to 2D-VRH system.

Table 4. A and n parameters obtained from fitting Seebeck coefficient data to $S(T) = AT^n$ model, in the temperature range between 100 and 200K.

Sample	Sr=0.05	Sr=0.10	Sr=0.15	Sr=0.20
A	21.6(1)	26.5(2)	31.2(1)	58.4(1)
n	0.50(1)	0.47(2)	0.36(1)	0.28(1)

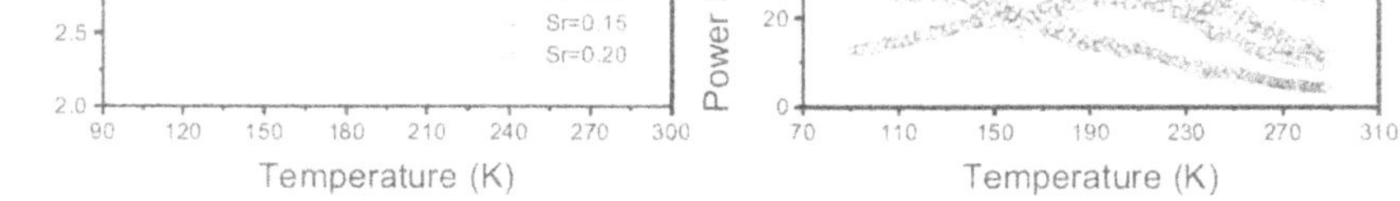

Figure 4. Temperature behavior of thermal conductivity [a] and [b] thermoelectric power factor $PF = S^2/\rho$ for LSCuO ceramics.

The electrical resistivity increases with increasing temperature showing a metallic-like behavior (see figure 3b). At room temperature, the magnitude of ρ(T) decreases with the strontium content from

5 to 0.4 mΩ-cm. This trend agrees with the observed behavior of Seebeck coefficient of these oxygen deficient LSCuO ceramics, which is perhaps related with the increase of carrier concentration as the Sr level increases.

The thermal conductivity is weak temperature dependent, its magnitude decreases with strontium level from 4.7 to 3.6 W/K-m (see figure 4a). The electronic contribution κ_e to total thermal conductivity κ ($\kappa=\kappa_e+\kappa_l$), was calculated from Wiedemann-Franz model ($\kappa_e = LT/\rho$, where is the Lorentz number for a Fermi gas, ρ the electrical resistivity and T the absolute temperature).

The electronic contribution to κ increases from 3.6% to 43% of the total thermal conductivity as the Sr level increases. Which could be a direct consequence of the observed behavior the grain structure which reduces the lattice contribution and the increase of the charge carriers density as Sr doping is increasing.

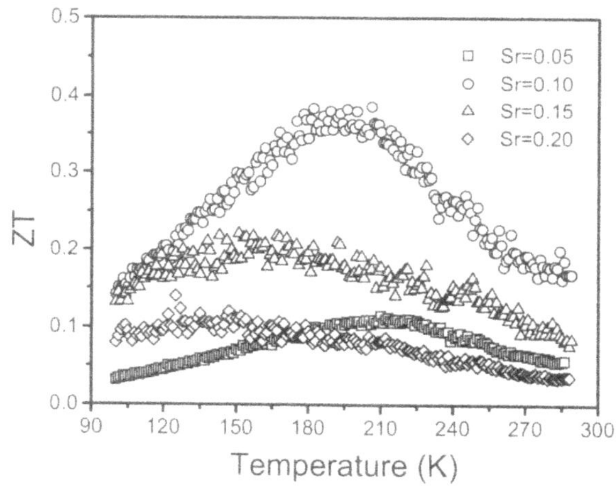

Figure 5. Temperature dependence of thermoelectric dimensionless figure of merit for LSCuO samples.

Similar results have been found in studies of thermoelectric properties of other oxide ceramics as $Ca_3Co_2O_6$, $LaCoO_3$, $HgBa_2Ca_2Cu_3O_{8+\delta}$, $HoBaCo_2O_{5.5}$ etc. which were widely studied by A. Maignan et al. [14-18] which confirm the coexistence of large Seebeck coefficient values with good metallicity, shown the potential applications of these oxides as thermoelectric elements.

The thermoelectric power factor PF, is a parameter that determines the electrical properties of a thermoelectric material; this performance parameter is defined as:

$$PF = \frac{S^2}{\rho} \tag{3}$$

where S and ρ are the Seebeck coefficient and the electrical resistivity, respectively.

Figure 4b shows the temperature behavior of power factor, PF increases with strontium doping up to maximum values close 80 μW/K^2-cm, in the temperature range between 170 and 200K for the samples with Sr=0.10.

The energy conversion performance of a thermoelectric device is evaluated using the dimensionless figure of merit ZT, which is a function of its transport properties, ZT is expressed by [19,20]

$$ZT = \frac{S^2 T}{\rho \kappa} \tag{4}$$

where S is the Seebeck coefficient, ρ the electrical resistivity, κ the total thermal conductivity and T the absolute temperature.

From S(T), ρ(T) and κ(T) data, the dimensionless figure of merit, ZT was determined the magnitude of ZT increases with the temperature up to reach values close to 0.4 in the temperature range between 170 and 210K, in the samples with Sr=0.10 (see figure 5).

The magnitude and the temperature behavior exhibited by PF and ZT suggest these LSCuO ceramics could be promising thermoelectric material, if they are submitted to proper changes in their chemical composition.

CONCLUSIONS

In summary, $La_{2-x}Sr_xCuO_{4-\delta}$ ($0<x\leq0.2$) polycrystalline samples were prepared by solid-state reaction method. The transport properties of the samples were modified by submitting the obtained samples to a post-growth thermal process at 900^{0}C during 50 hours. The samples exhibit a tetragonal crystalline structure and high Seebeck coefficient values, which decrease with Sr level. The electrical resistivity shows a metallic-like behavior and its magnitude is less than 5 mΩ-cm at room temperature. The magnitude of total thermal conductivity decreases with the Sr content, which agrees with the observed behavior in the grain structure. The studied transport properties yield maximum values for PF and ZT of 80 $\mu W/K^2$-cm and 0.40, respectively, which become this kind of ceramic compounds promising material for thermoelectric applications.

ACKNOWLEDGMENTS

The authors would like to thank to *"La división de Investigación de la Universidad Nacional de Colombia, Sede Bogotá"*, DIB for the financial support.

REFERENCES

1. I. Terasaki, Y. Sasago and K. Uchinokura, *Phys. Rev.B* ,**56**,12684-12689(1997).
2.Y. Yakabe, K. Kikuchi, I. Terasaki, Y. Sasago and K. Uchinokura,*Proc.16th Int. Conf. on Thermoelectrics*, 523-527(1997).
3. S. Li, R. Funahashi, I. Matsubara, K. Ueno and H. Yamada, *J. Mater. Chem.*,**9**,1659-1660(1999).
4. S. Yamanaka, H. Kobayashi and K. Kurosaki, *J.J. Alloys Comp.*,**349**,321-324(2003).
5. S. Li, R. Funahashi, I. Matsubara, H. Yamada, K. Ueno and M. Ikebe, *Ceramics International*, **27**, 321-324(2001).
6. E. Sudhakar, J.G. Noudem, S. Hebert and C. Goupil, *J.Phys. D: Appl. Phys.*,**38**,3751-3755(2005).
7. L.C. Moreno, D. Cadavid and J. E. Rodríguez, *Microelectronics Journal*, **39**,548-550(2008).
8. J.S. Kim, B.H. Kim, D.C. Kim and Y.W. Park, *Ann.Phys.* (Leipzig),**13**,43-47(2004).
9. A.L. Pope, B.Zawilski and T.M. Tritt, *Cryogenics*,**41**,725-731(2001).
10. *Thermal Conductivity, Theory, Properties and Applications*, Edited by Terry M. Tritt, Springer,New York, 2004.
11. J.F. Kwak, *Phys. Rev. **B***, **13**, 652-657(1976).
12. P.M. Chaikim and G. Beni, *Phys. Rev. **B***, **13**,647-651(1976).
13. N.F. Mott and E.A. Davis, *Electronic Processes in non-crystalline materials*, Claredon Press, Oxford, 1979.

14. S. Hébert, D. Flahaut, D. Pelloquin, C. Martin and A. Maignan, 22nd International Conference on Thermoelectrics 2003.
15. A. Maignan, S. Hébert, D. Pelloquin, C. Michael and J. Hejtmanek, *J. Appl. Phys.* 92, 1964(2002).
16. W. Kobayashi, S. Hébert, D. Polloquin, O. Pérez and A. Maignan, *Phys. Rev. B* 76, 245102(2007).
17. A.C. Masset, C. Michel, A. Maignan, C. Martin and B. Raveau, *Phys. Rev. B* 62,166(2000).
18. A. Maignan, V. Caignaert, B. Raveau, D. Khomshii and Sawatzky, *Phys. Rev. Lett.* 93, 026401 (2004).
19. D. M. Rowe, *CRC Handbook of thermoelectrics,* CRC Press, Boca Raton Fl, 1995.
20. G. S. Nolas, J. Sharp and H. J. Goldsmid, *Thermoelectrics, basic principles and new materials developments,* Springer-Verlag, Berlin,2001.

Author Index

www.ingramcontent.com/pod-product-compliance
Ingram Content Group UK Ltd.
Pitfield, Milton Keynes, MK11 3LW, UK
UKHW020820290926
15470UKWH00003B/1268

9 780470 905470